Hi-Fi
for the enthusiast

M L Gayford BSc, CEng, DIC

Hi-Fi
for the enthusiast

2nd (Enlarged) Edition

PITMAN PUBLISHING

First published 1971
Second edition 1975

00062076

SIR ISAAC PITMAN AND SONS LTD
Pitman House, 39 Parker Street, Kingsway, London, WC2B 5PB
PO Box 46038, Banda Street, Nairobi, Kenya

PITMAN PUBLISHING PTY LTD
Pitman House, 158 Bouverie Street, Carlton, Victoria 3053, Australia

PITMAN PUBLISHING CORPORATION
6 East 43rd Street, New York, NY 10017, USA

THE COPP CLARK PUBLISHING COMPANY
517 Wellington Street West, Toronto 135, Canada

490288782 PS

PP.

ISBN 0 273 00497 2

Reproduced and printed by photolithography and bound in
Great Britain at the Pitman Press, Bath
G 3361:14

Preface to first edition

The numerous people in many countries who are Hi-Fi enthusiasts aspire to enjoy the reproduction of music in their homes from equipment which not only offers better quality than that of normal commercial radio sets but also is suited to their individual needs as regards technical flexibility and the arrangement of the various elements in the home.

This book is intended for these enthusiasts and sets out to explain the most important technical factors in a straightforward though not over-simplified manner. The latest techniques used for pick-ups, turntables, amplifiers, tuners, tape-recorders and loud-speakers are highlighted and the important points to consider when buying or building units for home interconnection are stressed. Many pictures, diagrams and up-to-date references are included and some interesting and practical home-built systems of professional quality and appearance are illustrated.

I should point out, incidentally, that although this book has been written in good faith as a guide to the successful use of Hi-Fi equipment it is not implied that I can accept responsibility for unsatisfactory results! I should also emphasize that the fact that I have been given permission to reproduce certain designs and layouts does not mean that these are not protected by patents.

A wide variety of suitable units and components are now readily available and many home assemblers will find that the realization of a true Hi-Fi system need not overtax their skill or their pocket. The lasting satisfaction obtained will be their reward.

M L GAYFORD

Preface to second edition

More recently, there have been several developments in stereo-phonic sound reproduction which represent further steps in the endless pursuit of true and complete re-creation of musical sounds in the home.

Discrete four-channel stereophonic sound recordings have been introduced on magnetic-tape machines and cartridges which, when replayed on the proper good-quality reproducing sets, give two front and two rear channels for feeding to four loudspeakers.

Quadraphonic gramophone records of several different types have been produced, in which the recorded information on the two groove walls is ingeniously encoded so as to yield two front and two rear channels when replayed.

Again, there are now four-channel "synthesizer" circuits and ICs which will process any normal two-channel stereo programme by extracting the front difference signals and presenting them to the rear loudspeakers via a variety of circuits. The result is to add an enhanced reverberant "ambience" to the overall sound.

With such a wealth of material now available, the enthusiast must perforce experiment with more circuitry, more loudspeakers and different directional arrangements, if he is to decide which particular arrangements he prefers. Satisfaction in sound repro-duction is a very personal matter, and opinions will always differ.

To this end, two completely new sections have been added at the end of the book as well as additional references and some up-dating of the text.

M L GAYFORD
MARCH 1974

Acknowledgements

The author's cordial thanks are due to the following organizations and individuals for the provision of information and for permission to reproduce various illustrations (as indicated below).

Acoustical Manufacturing Co Ltd (Fig 41)
Antiference Ltd (Figs 6*a*, 7)
Audio Workshops Ltd (Fig 6*b*)
Briggs, Wharfedale Studio (Fig 4)
BSR Ltd (Fig 66)
Butterworth & Co (Publishers) Ltd (Figs 1, 2)
Connoisseur Ltd (Fig 16)
Dolby Laboratories Ltd (Fig 31)
Ferguson-Thorn Ltd
Garrard Engineering Ltd (Figs 15, 67)
Goldring Manufacturing Co (Great Britain) Ltd (Fig 20)
Grampian Reproducers Ltd (Fig 32*a*)
Hayden Laboratories Ltd and Nagra Ltd (Fig 26)
Heath (Gloucester) Ltd and Heathkit (Figs 8, 9)
Hi-Fi News
IPC Electrical-Electronic Press Ltd and C. W. Hellyer (Fig 25)
Imhofs-Bedco Ltd and H. J. Leak Ltd (Fig 68)
ITT Consumer Products (UK) Ltd (Fig 66)
ITT Europe Ltd (Figs 32*b*, 32*c*)
Jordan-Watts Ltd (Fig 40)
J W (Wholesale) Co (New Classic Storage System) (Fig 21)
KEF Electronics Ltd (Fig 45)
Largs Ltd (Fig 69)
Mechanical Copyright Protection Society (page 224)
E. W. Mortimer
Motorola Semiconductors Ltd (Figs 10, 64)

North East Audio Ltd (N.E.A.L.) (Figs 28, 30 and Table on page 99)

Philips Electrical Ltd (Fig 29)

Radford Acoustics Ltd (Figs 53, 54, 55, 56, 57)

Rank Wharfedale (Rank Radio International) (Fig 62)

Recordaway Storage Systems Ltd (Fig 22)

Rola Celestion Ltd (Figs 48, 49, 50, 51)

Sinclair Radionics Ltd (*Project 60 Manual*) (Figs 34, 35, 37, 38, 39)

SME Ltd and Thorens (Fig 14)

Staples & Co Ltd (Fig 70)

A. R. Sugden & Co (Engineers) Ltd (Connoisseur Equipment) (Fig 16)

Tannoy Ltd (Fig 52)

Toshiba Ltd and Erie Electronics Ltd (Fig 63)

John Walton (Fig 19)

Wilmslow Audio Ltd, Cheshire

Wireless World and P. J. Baxandall (Fig 36)

Wireless World, Geoffrey Shorter, Basil Lane and D. C. Read (Figs 59, 60)

Contents

The voice and other sound sources—The hearing process—Psycho-acoustical aspects of hearing—Stereophonic reproduction and binaural listening—Domestic stereophonic reproduction—Room acoustics: sound wave patterns—Loudspeakers in rooms—Stereophonic and quadraphonic sound

Radio tuners and sets—Aerial systems—Tuner principles—Typical high-quality FM tuner quantitative specification figures—FM stereo multiplex receiver units—Special characteristics of frequency modulation—FM stereo decoder circuits and specifications—High-quality AM tuner design—AF noise suppressors—Specification of a typical high-quality AM tuner—IC broadcast stereo decoders—Means of reducing noise in stereo radio reception—Whistle filters for stereo decoders—Quadraphonic broadcasting—FM inter-channel squelch circuits

Design and use of turntable and pick-up units—The manufacture of automatic record-changers and -players—The system as a whole—Properties of monophonic and stereophonic record grooves—The record groove—Stylus and arm angular effects—The record material and the stylus—Pick-up design and performance specifications—Stereo recording and reproduction—Checking procedure for correct pick-up tracking—Checking for stylus wear and damage—Pick-up arm design—Record care and cleaning

Magnetic recording tapes—Tape-recorder decks and machines—Volume-level indicators—Synchronization of tape-recorders and home cine equipment—Remote-control systems—Reel-to-reel stereo tape-recorders, cassette and cartridge machines—Cassette machine details—Dynamic noise-suppression systems—Dolby noise-reducing system details—The Philips dynamic noise limiter (DNL)

Illustrations

Hi-Fi for the enthusiast

Introduction

Music of all kinds has now become part of our lives. We have music while we work, music in the shops and music in the open air, and in the home we enjoy pop music, folk music, light music or serious music, according to our age and inclination. Never before has there been such a widespread desire for music and such a keen appreciation of it.

The production and reproduction of music now depends increasingly on electronic and electro-acoustical equipment and the technical standards of all types of system have risen enormously in the last few decades. Recording studios and broadcasters are equipped nowadays with expensive professional apparatus which is capable of very high standards of sound quality when handled with proper skill. Stereophony has added acoustic perspective to reproduced sound and, when skilfully used, gives a dramatic improvement in realism and emotional impact.

Contrary to opinions often voiced, jazz, modern pop and folk music benefit from high standards of sound reproduction as much as serious music. Complex rhythms, ingenious echo effects, vivid tone colours and vocalization can gain from high-quality reproduction as well as the more subtle but none-the-less compelling and dramatic performances of more serious or classical music.

Commercial radio sets and record-reproducers in the mass-produced cheaper field today can offer remarkable value for money in spite of enormously increased costs of production and distribution. Quality of sound reproduction and reliability are now both greatly improved as a result of many years of research and skilful engineering. Nevertheless, as is the case with most commodities, value is generally in proportion to the price and there are many people who are prepared to pay more to get increased enjoyment from a higher standard of sound reproduction.

An experienced Hi-Fi dealer is able to prescribe and install the best equipment for any individual need, but there is a considerable jump in price between mass-produced commercial sets and the best custom-built equipment. Many people, however, like to assemble "do-it-yourself" high-fidelity equipment from bought units, modules, kits or components, and it is here that great cost savings can be made without prejudicing satisfactory results, provided that the technical approach is along correct lines.

In this book it is proposed to suggest good outline designs and to stimulate ideas so as to enable you to avoid some of the pitfalls involved and to get good results from equipment which you can buy or make up. Although there are many excellent detailed designs published for amplifiers, tuners etc, both in the technical periodicals and by manufacturers of kits, there is a dearth of advice on the satisfactory integration of equipment units and in design of the cabinets and built-in enclosures which are so essential to the proper use and enjoyment of high-fidelity systems. It is here, particularly, that you can produce attractive individual arrangements to suit your own taste and domestic layout without being involved in exorbitant expense.

Television is also an established part of home entertainment and should be considered when a comprehensive entertainment assemblage is being planned. Television sound and quality is potentially of a very high standard and is worth feeding to a high-quality audio system.

The acoustic properties of listening rooms is of considerable importance, particularly for stereo reproduction, and a great deal can be done by comparatively simple acoustic treatment and rearrangement of existing rooms. If you are considering building a new house it is desirable, so far as possible, to design your listening room for the optimum properties.

In this book we will deal primarily with sound reproduction in the home or in small halls, together with aspects of sound recording which may also be of interest to those who wish to produce their own programmes. Recording should logically be treated first, but it is not of primary interest to many people who are building equipment, and so it is dealt with in a subsidiary manner.

We will consider first the sources of programme available,

the equipment concerned being radio and TV tuners, record-players, tape-recorders and replay units, film sound-tracks etc. Amplifiers, control units, equalizers and filters of both valve and semi-conductor types are next dealt with, and then cabinets, housings and loudspeakers of various kinds, including multi-unit assemblies using electrical cross-over networks.

Before going into detailed consideration of the various parts of the system, however, we will deal briefly with the nature of speech and hearing, the acoustics of rooms and buildings, and the way in which loudspeakers may be used to produce the right sound.

Sound and hearing

The voice and other sound sources

When we inquire into the art of sound reproduction we must begin by studying the nature of human speech and hearing, as these are the basic elements involved. The making and appreciation of music came much later in human evolution but are now of equal if not greater importance as far as the high-quality reproduction of sound is concerned.

The first sound source with which we are concerned is therefore the human voice, as used in speaking or singing. It is an extremely complex mechanism, as regards both the rapidly varying waveforms produced and the complicated form in which the sound is radiated from the mouth, nostrils and, to a certain extent, from the walls of the throat and chest. The accurate re-creation of the speaking or singing voice requires, in the first place, an understanding of the mechanism in order that correct microphone and loudspeaker design and placement techniques may be employed.

Physically, the power of the human voice is produced by the air flow from the lungs, which is expelled by the action of the muscles of the diaphragm and rib cage. Fig 1 shows the parts involved in speech and clarifies the dominant modes of operation of the various parts of the throat, mouth etc. Vocal utterances are broadly recognized as "voiced" and "unvoiced" sounds. "Voiced" or semi-continuous waveforms are composed of discrete frequency components, used for vowels or a sung vocal output. These are generated by the air-flow modulated by vibrational opening and closing of the glottal vocal slot, producing the fundamental voice frequency, the harmonics or formants being augmented by the natural resonances of the other parts of the vocal tract, represented by the various cavities and constrictions. "Unvoiced" sounds such as "plosive" transients or hissing sibilants are produced by either closing off or releasing the air-flow abruptly by the tongue, lips etc, or by causing turbulent "edge tones" etc by forcing the air-flow through narrow constrictions formed as required, for example, by the tip of the tongue against the teeth.

In fact, the character of a person's voice and much of the basic intelligence or substance of the vocal message are conveyed by

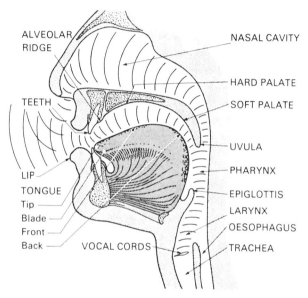

Fig 1. *A representation of the speech organs showing the vocal-chord sound-source, the formant-producing vocal tracts and the radiating mouth and nasal outlets.*

quasi-continuous utterances of near syllabic length (ie appreciable fractions of a second or a greater length of time). The distinctness or articulation is largely due to the unvoiced, transient, or sibilant sounds.

The range of variation of the human voice with regard to intensity, frequency range and emotional impact contributed by variations in stress, pace of utterance, pitch etc is enormous and it is not surprising that near-perfect reproduction of the human voice represents a searching test of a sound system. In fact, ideally a dynamic intensity range of 60 dB (1000 to 1 pressure ratio) and a frequency response from about 50 Hz to nearly

20 kHz are required, though it is true that it is possible to transmit intelligible, as opposed to completely natural, speech with a mere fraction of these values, as represented by the parameters of many systems designed to provide basic communication rather than the highest quality.

In order to pick up the various sounds produced by a human voice in the correct manner, a microphone must be designed to accept sounds from the mouth, nose etc accurately, without being influenced by the unidirectional blasts of breath which inevitably accompany speech, although these do not form part of direct sound when heard at the normal listening distance of a few feet. The microphone must also be placed so as to obtain the correct "acoustic perspective" for the voice in the particular room or surroundings where speaking is taking place. Broadly, this means that the microphone must not be so close as to react directly on the characteristics of the voice, by presenting an obstacle near to the mouth opening, nor must it be so far away as to pick up an excessive amount of reflected or reverberant room sound. In this connection, even a fairly directional microphone may not approach the remarkable discriminating power shown by the human auditory system when listening directly with two ears under comparable conditions.

Musical instruments of various kinds, on the other hand, differ considerably from the human voice and from each other, each type tending to have its own peculiarities. In general, they represent larger and more powerful sound sources than a voice, possessing horns, sounding boards, resonators and other sound volume-increasing devices. The waveforms of music are generally more constant and more continuous but there may still be important transients. Instruments often exhibit more markedly directional properties than a voice and this affects the required microphone placement techniques. Instruments may also produce subsidiary unidirectional blasts of air and other minor effects such as blowing, or mechanical operating sounds due to pedals, hammers, valves etc. These may also be picked up and exaggerated by an excessively close microphone.

Further, the whole question of stereophonic reproduction involves, in effect, another dimension in sound reproduction and may require complicated microphone techniques to produce a good subjective stereophonic effect when reproduced by 2/4 channels and finally listened to by two ears in the room where the sound is reproduced. It is important to distinguish between inter-channel differences existing in the necessarily artificial two-channel stereo system and the inter-aural effects existing around and within the head of the human listener. Two-channel electronic stereophonic reproduction is, in fact, an artifice which produces an illusion of sound source directionality and position by providing the human auditory system with enough of the right kind of clues.

The hearing process

The human auditory system is even more complex than the vocal system. The ear consists of an outer area, in which the head, the pinna and the ear canal are all important acoustically,

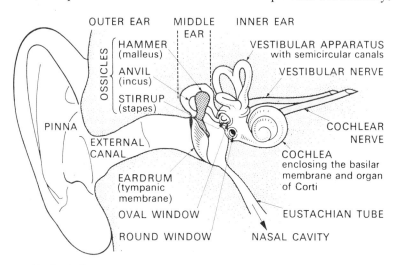

Fig 2. A diagrammatic view of the human ear which gives some idea of the complexity involved. The higher co-ordinating functions of the brain involved in the auditory functions are even more complicated and less well understood.

the middle ear system, the inner ear and neural system and, finally, the cortical and brain functions at various levels which are concerned with all aspects of the hearing process. The various parts of the ear are illustrated in Fig 2.

This shows a section through one complete ear and reveals the outer, middle and inner ear structure. The acoustic obstacle effect of the head modifies the response of each ear above about

1000 Hz, increasing the response of the ear to sound from the same side by up to 6 dB and reducing the response of that ear to sound from the opposite side. This difference in response with angle plays a part in directional hearing and also means that there is a difference between the pressure and the free-field response of the ear, as with a sound-pressure-operated microphone. Thus an earphone which is made to generate constant sound pressure in the ear canal will not produce the same sensation at all frequencies as will a freely incident plane sound wave which has access to the complete head, owing to the difference in response.

The pinna or convoluted outer ear structure protects the ear canal entry and supplies some directionality and a resonant cavity gain at middle-to-high frequencies for free-field listening.

The ear canal (meatus) is about 2·7 cm long and 0·7 cm diameter. It is terminated by the tympanic membrane or eardrum, which is an inwardly-directed cone of about 0·8 cm^2 in area. The first acoustic resonance of the outer ear and canal occurs at about 3000 Hz and provides a sound pressure gain at the eardrum. The adjacent middle ear cavity contains air (equalized to atmospheric pressure via the Eustachian tube) and the ossicular bone linkage system. In addition to providing mechanical coupling between the eardrum and the cochlea-coupled oval window, the ossicles provide an impedance transformation by lever action.

It may be noted that an earphone placed externally on the pinna is coupled to the ear canal in a complicated way; firstly because the individual convolutions of the pinna (which vary from person to person) represent a series of communicating passages and cavities feeding into the ear canal and, secondly, because a normal hard substantially flat ear-cap does not represent a constant sealed connection to the pinna, but one which usually involves a substantial leak to the outer air, the exact amount of leak depending on the way in which the earphone seats on the ear and the force which is applied to flatten the pinna. This leak can cause a serious loss of lower frequencies below 1000 Hz in the earphone response, whilst the acoustical passages can seriously modify the higher frequency response above 1000 Hz. These latter effects are even more severe in the case of large circum-aural ear-pad muffs, which are often provided in order to give better wearing comfort and improved noise exclusion.

The need for the maximum coupling efficiency between an earphone and the ear and the desire to eliminate external noise has led to the development of the insert-type earphones where a small tube or boss is fitted into the ear canal entry so as to make a seal. The small size of such transducers has tended to restrict the low-frequency power available and thus they have been largely used for the hearing aids and general communications rather than for high-quality reproduction.

If the ossicles become incapable of motion through disease, conduction deafness results, making air-path hearing aids useless. Bone-conduction receivers can be used to overcome this disability.

The inner ear includes the fluid-filled cochlea spiral, of length about 35 mm in all. The cochlea lengthwise partition consists of two discrete spaced membranes, one of which is the tapering basilar membrane with which the organ of Corti is in contact. The latter consists of many thousands of sensory cells terminating the auditory nerve. The inner and outer hair cells are among these and are in contact with the tectorial membrane. Deflection of the basilar membrane, due to a travelling wave excited by the stapes at the oval window, causes relative motion between itself and the tectorial membrane. This in turn deflects some of the hairs and produces electro-neural discharges in the associated nerves. The basilar membrane consists of tapered distributed elements and constitutes a non-reflecting dispersive medium. The amplitude and phase response of the various points along the basilar membrane simulate a succession of tuned filters of approximately constant sharpness or Q, linear increments of length along the membrane corresponding roughly to logarithmic frequency intervals, at least for middle and low frequencies. Thus a given frequency excites a maximum response at the appropriate distance along the membrane. The latter is therefore a frequency-analysing device with rather broad-band tuning. The reason for the extremely fine increments of pitch which can be discriminated must be sought in the higher levels of neurological and mental auditory processing.

Temporary deafness due to loud sounds is thought to be due to evanescent chemical changes in the hair cells. Prolonged exposures may cause permanent deafness, due to damage, and irreversible changes in the hair cells. Thus prolonged listening to music or other sounds of excessive loudness must be indulged in with caution in order to avoid auditory damage.

The cochlea fluid is subjected to compressional waves via the stapes and oval window, the fluid pressure waves ultimately being relieved by motion of the membrane covering the round window, which is in contact with the fluid on the remote side of the membrane complex which divides the cochlea spiral down its length. The vestibular apparatus and semicircular canals which act as a balance-sensing system communicate with the upper end of the cochlea above the oval window.

The physiological and mechanical functioning of the outer and middle ear and also the electrical neural potentials produced in the cochlea have been established and explained by G. Békésy in his prize-winning work,[1]* and by many other workers. The exact functions occurring in the brain in all the complex sensing, correlating, processing, association and memory activities which are concerned in the various aspects of our understanding, appreciation and use of the sounds which we hear are still largely a matter for speculation.

The basilar membrane thus performs only a rough kind of frequency analysis, whilst the middle-ear muscles perform an automatic intensity-limiting function by reducing the mechanical coupling via the ossicular bones. The intensity of the perceived sound is transmitted to the brain in the form of electrical coded nerve pulses, the number of nerves "fired" in a given bundle and the pulse repetition rate conveying the information. Thus we see that auditory neural signals resemble a form of coded "digital" transmission rather than the "analogue" electrical waveforms with which we are familiar in sound-reproducing systems.

The brain is able to analyse the sounds incident on each ear, as regards frequency and intensity, with a remarkably fine resolving power. It is also able to conduct a running correlation on the signals from each ear to obtain directional sensations with regard to the incident sounds. The further running analysis of the complex and rapidly varying waveforms of speech alone and the comparison with previously stored memories of earlier sounds, together with all the other mental functions associated with the understanding and appreciation of sound is scarcely short of miraculous when compared to the common mechanisms with which we are more familiar.

* Superscript numbers in parentheses refer to the list of books and references on pages 225–6.

Psycho-acoustical aspects of hearing

We have considered broadly the way in which the human speech and hearing system functions, apart from the mental processes involved, the precise working of which can only be guessed at. However, a great deal of experimental work has been done to systematize the exact way in which we react to sounds of various types. It is desirable to have a knowledge of these characteristics in order to design and use high-quality sound-reproducing equipment.

The main aspects under which the psycho-acoustical hearing characteristics are studied are—

1 The perception of pitch (frequency).
2 The perception of intensity (sound pressure).
3 The appreciation of time duration, both in regard to transient sounds and in the recognition of small time intervals, particularly with regard to the arrival timing of wavefronts and the spatial localization of sound sources.
4 In the subjective impression we receive of the loudness of complex sounds, which depends on whether the sounds are impulsive or continuous in time and also on the frequency spectrum of the sound, as opposed to a simple objective measurement of the average or peak acoustic pressure.

Of these various psycho-acoustical factors, apart from those affecting the fixed performance parameters of the reproducing equipment, such as the basic frequency response and transient response, the main factor where adjustment may be required is the "scaling" of the shape of the frequency response when the reproduced level differs greatly from the original sound level at the microphone position. The response curve contours change with the incident sound pressure level, the most marked feature being the loss of sensitivity to low frequencies. This phenomenon is particularly noticeable when quiet male speech is reproduced at an unnaturally high level and gives rise to the so-called "loudness control" on some amplifiers, whereby the bass is automatically adjusted in level as the volume control is turned up. Unfortunately, the design of this type of control is complicated by the different low-frequency characteristics of speech and music.

Other types of "shaping" of the amplifier response curve are aimed mainly at minimizing various unwanted noises, such as low-frequency rumbles, hum etc, and high-frequency electronic noise, tape or disc noise etc, as well as higher-order distortion products. Some "equalization" or correction of known deficiencies in transducers etc is also used.

Stereophonic reproduction and binaural listening

It has been known for a long time that more than one reproducing channel is required if sound is to be reproduced in another place in a really satisfactory manner. Until recently economic considerations have limited nearly all communication systems to a single channel.

Thus the telephone is restricted to a single link, broadcasting is still largely performed through a single-channel transmitter, and a monophonic gramophone record is endowed with a groove executing lateral excursions only.

Nature has provided us with two ears and the nerve impulses from both ears are correlated by the brain so as to give definite subjective impressions as a result of almost any given acoustic stimulus, however slight. It is natural for us to listen with two ears to sounds approaching from various directions and we are able not only to form an accurate impression of the direction and character of sound sources but also to discriminate in a remarkable fashion in favour of wanted as against unwanted sounds.

Thus a monophonic reproduction of anything but the simplest sound source in idealized surroundings is a somewhat unnatural experience and we are unable to exercise our full psycho-acoustical powers. In many cases, the slight subconscious disturbance which is likely to accompany any unnatural experience would appear to be likely to prevent our full enjoyment or appreciation of monophonic reproduction of sounds, such as music, which have an aesthetic appeal. It is also common experience that we are to some extent unable to discriminate against unwanted background interfering noises or excessive reverberation when listening to normal monophonic sound reproduction, eg radio speech etc.

Present-day stereophonic sound reproduction might thus be partly defined as an attempt to provide extra directional information so as to make the result more natural and so to enable more normal psycho-acoustical processes to function for the listener. Economic factors limit us to 2/4 channels for domestic stereophony. More channels enable better stereophonic effects to be obtained but the gain in realism obtainable with two channels as compared to one can still be very marked.

To obtain the full benefit of binaural listening it is not sufficient to feed separate channels to each ear by earphones. The ears must be immersed in the "free" sound field of a room and the head must be free to turn and to make other subconscious movements which are vital to sound localization and correct natural auditory perception. Binaural listening on headphones can, however, give many of the benefits of stereophonic reproduction, although there will be certain shortcomings.

Early stereophonic experiments used a number of microphones spaced in a line across the front of a sound stage. A corresponding line of loudspeakers in another auditorium produced an approach to a "curtain of sound." It was found that three separate channels gave a fairly good stereophonic effect as regards naturalness of reproduction and spatial localization of sounds.

The three channels were carefully balanced to have identical frequency and phase-delay characteristics. When a sound source is nearer to one microphone than to the others, two effects operate. Firstly, the direct sound intensity will be higher on the corresponding loudspeaker and, secondly, the sound will emanate from it sooner than from the other loudspeakers by a time dependent on the relative microphone distances. The velocity of sound-wave propagation is such that a path difference of one foot introduces a relative time delay of one millisecond. Thus we see that the stereophonic effect depends both on interchannel intensity differences and on interchannel time differences. It will also be noticed in a successful stereophonic system that a sense of spaciousness and direction is imparted to the reflected or reverberant sounds picked up by the microphones as well as to the direct sound.

In order to get good stereophonic reproduction over a large-audience seating area in a fair-sized hall or cinema it has been found necessary to use more elaborate systems utilizing a larger number of channels. Good results may be obtainable on or near the centre axis of the loudspeaker system when a limited number of channels is used, but a loss of stereophonic illusion and a defect described as "concaving" or "hole in the centre" are liable to be observed at some off-axis seats. The latter defect is manifested when a sound source moves across the stage. The source appears to recede or follow a concave path between the loudspeakers.

Five-channel stereophonic film systems using six magnetic

sound tracks at the sides of 70-mm film have been used to give good stereophonic effects in cinemas. Five loudspeakers are spaced across the stage, the remaining channel being used for extra sound effects, produced in some cases by loudspeakers at the rear of the auditorium. This is done in order to get over some of the short-comings of a stereophonic loudspeaker array which originates sound in a line in front of the listener. A few additional loud-speakers at suitable places can be erected as required and fed by an extra sound-track. Also, the full natural effect of reverberant sound cannot be reproduced unless some loudspeakers are distributed about the auditorium. Generally these must be fed with a signal having a deliberately introduced time delay as compared to the frontal loudspeakers, when used to enhance the reverberant quality. No artificial delay is added to the signal fed to the supplementary loudspeakers when it is desired to give the impression of a direct sound to the rear or above the audience. In this case, a common delay might be added to all the frontal stereophonic channels. The effect whereby the sound appears to originate from the source which originates the first sound to arrive at the listener was first investigated and analysed by de Haas.[2]

At about the same time as the early three-channel experiments were carried out, A. D. Blumlein of EMI proposed a different form of stereophonic microphone technique. In place of microphones spaced apart along a line or arc, coincident directional micro-phones were used with their major lobes of polar response oriented so as to point respectively to the left and right of the sound stage. The results were originally applicable only to a two-channel stereo-phonic system. No significant time differences occur between the microphones, as these are mounted as close together as possible. A sound source at an angle to the centre axis produces different amplitudes in the two channels as a result of the particular shape of the directional polar characteristics of the microphones used. It might be thought that the lack of inter-channel time difference information would seriously reduce the stereophonic effect obtainable. However, an effective amplitude-to-time conversion occurs at the listener's head due to the finite spacing between his ears.

The spaced microphone system is prone to some spurious effects, particularly when moving sources are involved. For example, the "Doppler effect" due to a rapidly moving source passing spaced microphones is reproduced at each microphone in turn, thus giving rise to several virtual sources when the sounds are reproduced. The crossed microphone system is free from this type of defect, though it has other limitations of its own. Thus it is sometimes difficult to obtain a proper stereophonic effect from sources spread across a wide sound stage unless the crossed microphones are moved well away from the stage, so giving an excess of reverberant sound. This follows from the fact that the entire stage width must be covered within the $\pm 45°$ angle between the main lobes of the crossed directional microphones. Any sources subtending angles greater than $\pm 45°$ may be reproduced at a reduced intensity on one channel and with a phase reversal on the other. In the case of crossed bidirectional (figure-of-eight) microphones, reverberant and other sounds emanating from behind the microphones will be reproduced with a $180°$ change of phase (an inherent property of a simple gradient-operated microphone). This may give the reproduction a disturbing or unnatural quality. Crossed unidirectional or cardioid microphones are somewhat better in these respects. It is not, of course, possible to have two distinct crossed microphones at exactly the same point in space. They are commonly mounted closely one above the other, often in the same casing. It has been found that the necessary finite vertical displacement between the microphones may introduce a small but sometimes noticeable inter-channel time difference for sounds emanating appreciably above or below the centre axis of the microphones, which may be interpreted by anyone listening to the final two-channel stereophonic reproduction as a horizontal displacement. Thus a person's voice and his footsteps might appear to come from different parts of the stage. A small horizontal displacement between the microphones is therefore considered preferable. In the case of very small capacitor microphone capsules, the vertical spacing can be made so small as to be negligible. Nevertheless, simple crossed ribbon or cardioid microphones will often give excellent amateur stereo recordings.

In practice, when an ambitious stereophonic recording is being made of, say, a large orchestral, choral or other performance involving a wide sound stage, it is possible to resort to a mixture of crossed and spaced microphone techniques.

Domestic stereophonic reproduction

Most producers of stereophonic programmes intended for domestic reproduction assume that the listener will sit near the centre axis of two loudspeakers spaced about 10 ft apart in a reasonably well-furnished living room of moderate size, the listening position being about in the centre of the room. It is advisable to have some acoustic absorption (eg curtains etc) in the side walls and on the wall behind the loudspeakers so as to minimize any loss of stereophonic positional accuracy due to first reflections of sound from these walls. Fig 3 (page 17) shows a plan view of such a room. It is possible that two loudspeakers with a fairly broad polar sound distribution are less critical as regards room conditions generally, but an improvement in definition and better stereophonic positional accuracy may be obtained if two loudspeakers with an optimum degree of directionality are used, eg two bidirectional electrostatic units or other loudspeakers with a carefully controlled higher frequency distribution.

The present means of general presentation and distribution of stereophonic programmes are by radio broadcasting (originally by two separate transmissions but now by multiplex modulation of a VHF transmitter), by dual-track tape-recorder or by stereophonic gramophone records, cassettes, cartridges.

Recently, attempts have been made to carry more directional information on the two tracks by making further use of the instantaneous relationship between the sum and difference signals in the left and right channels. Extra derived signals have then been fed to subsidiary front and rear loudspeakers in order to give a better acoustic ambience or sense of direction to the reverberant sound. The name "quadraphony" has been given to some of these developments. (See matrixing techniques, p. 197.)

Room acoustics: sound wave patterns

The sound pattern set up in a room is markedly different from the effects of sound propagation in the open air. Any source of sound causes alternate regions of increased and reduced pressure to travel outwards from its emitting surface. The source may be a pressure generator, ie a device which introduces alternating "puffs" of air under pressure into the surrounding air. An organ pipe, most "blown" musical instruments and certain specialized types of loudspeaker, eg the ionophone[16] or the pneumatic "loud-hailer," are examples of this type of sound generator. However, a more common type of source is a vibrating surface of a size sufficient to move enough air particles in contact with it to radiate an appreciable pressure wave. The normal domestic loudspeaker is of this type, in common with most stringed or percussion musical instruments. The human voice is, of course, a pressure type of source.

A sound wave started in the open air spreads out indefinitely, only suffering reflections from the ground, buildings, hills etc. Wind or thermal air currents will bend the wavefront, causing increased or decreased attenuation in the line of propagation, according to which way the wavefront is bent. For example, sound propagated with the wind is bent towards the earth and attenuation is reduced, ie the sound "carries better." The exact opposite occurs for sound propagated against the wind.

On the other hand, a sound wave propagated in a room is continuously modified in a very complex manner. It is reflected from all the boundary surfaces of the room and is selectively absorbed at each reflection, to an extent dependent on the porosity or other energy-absorbing properties of the surfaces.

The wavefronts are scattered or "diffused" by the presence of solid objects in the room or irregularities on the walls or ceiling. Some objects or surfaces in the room will be set into vibration by the sound waves and, if they are strongly resonant, their vibrations may persist long enough for them to be audible as subsidiary sound sources. Where walls, floors, ceilings etc, vibrate appreciably the effects may be described as structural vibrations of the room or building.

All these phenomena are studied in more detail in works dealing with room acoustics.[2,3] Note in particular, sound waves in a room sooner or later in their progress get into certain established "tracks" or room-resonant modes, which occur at various frequencies in one, two or three dimensions.

The overall effect on human hearing of these complex soundwave phenomena is the crux of our problem. We have seen that the human hearing process may be thought of as having three main functional processes. The physical outer and inner ear mechanism may be regarded as an acoustic transducer of great efficiency and sensitivity, which responds to sound pressure variations and transmits equivalent nerve pulses to the brain. This in turn may be regarded as producing a running display of the instantaneous sound patterns at each ear. The human mind may for this purpose be regarded as a conscious scanning system associated with a memory store. The memory, built up by heredity and a lifetime of experience, enables the mind to carry out a continuous correlation, recognition and appreciation process on the sounds incident on the two ears. Thus, for example, the direction of a sound source may be almost instantaneously estimated, primarily as a result of the slight time difference taken by an oblique sound wave to travel around the head.

For present purposes, it is sufficient to note that we are able to assess the rapidly changing sound pattern is a room with extraordinary accuracy. For example, we are still able to locate a small sound source in a far part of the room, in spite of the fact that the direct sound from the source is to all intents and purposes swamped by a multitude of sound reflections from the walls. This is due to the "precedence effect." This means in general that if the direct sound arrives a few thousandths of a second (milliseconds) in advance of the wave reflections from the walls we are able to identify it and to locate the source, as long as we are listening with two ears. This power is lost when a normal monaural single channel is used, eg by putting a microphone in the room and listening at some other place via a loudspeaker. Stereophonic reproduction via two microphones and two loudspeakers can go a long way towards restoring the directional discrimination lost in monaural sound reproduction.

Loudspeakers in rooms

We see that the human speech and hearing processes are extremely complicated and that, strictly, it is necessary to study them in some depth in order to appreciate the factors involved in the use of loudspeakers in rooms. The complicated effects of room acoustics are also involved here. We wish not only to re-create speech and singing but also to reproduce realistically musical works by large numbers of instruments. When one considers the disparity between the many sound sources and the large and complicated waveforms set up by a symphony orchestra in a concert hall and the vibrations produced by a paper or plastic loudspeaker cone in a living room, some idea of the difficulties involved is brought home to one.

The heart of a loudspeaker is the transducer or motor unit. On its own it is unable to produce effective sound waves at low frequencies; it requires either a large radiating surface, as in the electrostatic loudspeaker, or some form of baffle or cabinet. Two distinct basic types of loudspeaker exist. In the first type, the rear of the unit is completely closed off by a sealed cabinet or box, so that only the front side of the cone is able to radiate sound into the room. In all other types of loudspeaker the rear side of the diaphragm, which is generating pressure waves in anti-phase to those at the front, is connected to the air in the room either directly, in the case of a flat baffle or a shallow open-backed cabinet, or via the port or other radiating openings in the cabinet walls.

The manner in which sound is radiated from a surface or an opening depends very much on the frequency. At the lowest audible frequencies the wavelength of sound is about 30 ft, which means that our sound sources are so small as to constitute omnidirectional radiators of spherical waves. An enclosed cabinet loudspeaker consists of one such source, whilst the open type represents two point sources, one behind the other working in opposition. These tend to neutralize each other and to produce less bass. They also have a bidirectional response, in that there is no resultant output in the plane of the real or imaginary baffle between them, producing a figure-of-eight polar curve similar in shape to that of a ribbon microphone. Thus we see that the purpose of a port or acoustic labyrinth must be to reverse the phase of the back radiation so as to aid that from the front, and to maintain the bass response.

At extreme high frequencies the wavelength of sound is less than one inch, which is considerably smaller than the average loudspeaker cone. The effect here is that, instead of being omni-directional, sound is radiated in the form of directional beams. Where more than one sound source exists in near proximity, the effect is that interferences are produced at intermediate angles in particular, causing an irregular response.

We can summarize two of the main problems of loudspeaker reproduction as, firstly, maintaining a reasonably good efficiency at low frequencies and, secondly, obtaining a fairly consistent polar response, that is, a good omnidirectional or bidirectional distribution of sound at middle and low frequencies, together with a sufficiently well-spread and even polar sound distribution at higher frequencies.

There still remains the problem of how best to place the loud-speakers in your room in relation to the listening position.

No two rooms are ever exactly the same as regards their size, shape or acoustic properties, and the quality of the sound repro-duction may depend to a large extent on the siting of the loud-speakers. A number of theories exist as to the ideal arrangements for various given cases. One is that most loudspeakers act as sound-pressure generators at middle and low frequencies and that if they are mounted in a corner, the converging walls act as an extension of the loudspeaker and increase the low-frequency response. Also, a room has a large number of natural resonant modes of the air and most of these have maximum pressure points near to the corners. Thus the maximum number of these natural resonances should be excited by a pressure source in the corner, resulting in a smoother response of the reverberant sound in the room.

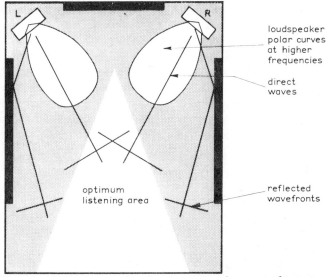

Fig 3. *An idealized room plan showing corner placement of two stereo loudspeakers in a symmetrical arrangement. Only two reflected waves are shown. In practice an almost infinite number of reflections blend to make up the reverberant sound. Excessive reverberation may "colour" the sound reproduction and can blur the stereo images.*

In practice the natural resonances in a room are not always evenly spread and structural vibrations of plaster-boards, floors and other large areas may exist. Undesirable resonances may be excited and so in practice it may be necessary to experiment to find the best loudspeaker positions, having regard to the limitations imposed by the doors, windows, fireplace and furnishings. The height of the loudspeakers should be such that the listener's ears are not too far off the centre axis of the tweeters or the main cone in the case of single diaphragm loudspeakers. It is here that bookshelf loudspeakers rather than floor-mounted cabinets may have an advantage.

For good stereophonic reproduction, too, one must obviously attempt to set a symmetrical "sound stage" about a central listening area in the room. This means that the two loudspeakers must usually be about ten feet apart, equally spaced with respect to the side walls, which should preferably be symmetrical as regards their acoustic absorbing or reflecting properties. Most living rooms cannot be re-planned especially for "stereo listening" but, fortunately, we can appreciate many of the benefits of stereo even if some of the exactitude of sound image positioning is lost.

Stereophonic and quadraphonic sound

We have seen that most loudspeakers tend to be directional to some extent at higher frequencies. Well-designed systems are arranged to maintain a fairly constant directional polar curve at the important mid-high and high frequencies. Some loudspeakers are now made to be almost completely omnidirectional at all frequencies in the horizontal plane, and even vertically as well, being almost spherical radiators. This usually requires the use of a number of small multi-facing mid-range and tweeter units. Either type may be used for stereophonic sound reproduction. Listeners are divided in their preference for one or the other arrangement, particularly now that four-channel or quadraphonic stereophonic programmes are available.

The preferred type of loudspeaker directional characteristics for any given room or musical tastes may well be a matter for experiment. Loudspeakers which are somewhat directional, as in Fig 3, are often found to give a good stereo positioning effect. Turning the loudspeakers more inwards, so their direct wave axes cross more in front of the listening area shown, is sometimes found to give a wider zone for good stereo effects.

On the other hand, it may be found that the general stereo effect and the ambience are enhanced if omnidirectional loudspeakers are used for frontal stereo, as well as for the rear positions when four-channel stereo programmes are reproduced. Figure 48 on page 166 shows front and rear loudspeakers arranged around a room. The positions may, of course, be varied to suit any set-up.

Loudspeakers of both types are described in chapters 7 and 8 of this book.

2

Programme input sources

In the previous chapter we have briefly surveyed some of the overall considerations which concern our hearing and the reproduction of music on a small scale. We will now proceed to more detailed consideration of the individual parts of the sound-reproduction chain, beginning with radio input units.

Radio tuners and sets

Sound broadcasting is carried out on the short, medium and long wavebands, in addition to VHF (very-high-frequency) bands. Television radiations are on the VHF and also the UHF (ultra-high-frequency) bands. A very large number of stations now operate in most countries on officially allocated wavelengths. In addition there are sometimes some unofficial transmitters.

Some important factors that militate against high-quality broadcast reception are—

1 Interference from other stations (programme break-through, "hash," carrier beat whistles etc).
2 Noise interference (crackles, man-made and natural static etc).
3 Distortion (modulation and demodulation distortion, sideband cutting, selective fading, multi-path distortion).
4 Tuning drift and signal variations (circuit tuning drift, multi-path propagation fading, aircraft reflection flutter).

Some of these troubles can be dealt with by sophisticated receiver design and by the use of elaborate aerial (antenna) systems, but it unfortunately remains true that for the best results a fairly high signal strength from the wanted station is usually necessary. This implies a powerful transmitter of fairly close proximity to the station, the preferred distance depending to a fair degree on the frequency.

The relatively close frequency spacing of stations in the long, medium and short wavebands and the long range of interfering transmissions makes it difficult to obtain true high-fidelity reproduction on these wavebands. The transmissions have perforce also to be amplitude-modulated (AM). This type of modulation makes it difficult for deeply modulated signals to be rectified without the introduction of some intermodulation distortion.

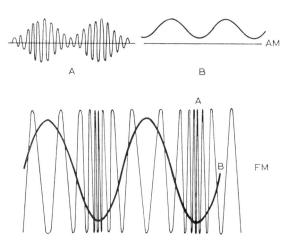

Fig 4. *Amplitude modulation and frequency modulation of a radio carrier wave.* A, *Modulated carrier.* B, *Audio-frequency signal received.*

(After Briggs and Roberts, *Aerial Handbook*)

The much greater "ether space" available in the VHF broadcasting bands and the limited range of transmissions at these frequencies enables frequency modulation (FM) to be used. This allows distortion-free demodulation of relatively loud audio signals to be carried out. As the programme matter is carried entirely by changes to the instantaneous frequency of the radio frequency carrier, interference can be greatly reduced. Most man-made and natural interfering radio signals are manifested as variations in the carrier amplitude; thus an efficient limiter in an FM receiver can cut out or greatly reduce most of these types of interference. Multi-path varying signals due to reflections of the radio wave from aircraft remain, however, a serious and indeed a growing trouble with FM.

The detailed design of good AM and FM tuners should in-corporate various circuit refinements which help to obtain the best results. These will be considered later.

An audio-frequency feed can of course be taken from most radio and TV sets by using the loudspeaker outlet connections, but this is not always the best way of obtaining a high-quality sound output from a set. Sometimes better results may be obtained

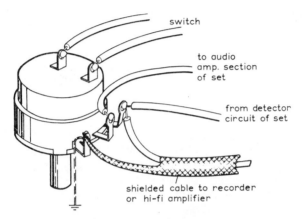

switch

to audio
amp. section
of set

from detector
circuit of set

shielded cable to recorder
or hi-fi amplifier

Fig 5. A connection taken from the audio-volume control of a commercial radio set in order to avoid the distortion which may be introduced in the output stages at higher signal levels. A suitable iso-lating transformer must be used for TV and radio sets used on mains without a mains transformer (ie normal "transformerless" sets) in which the chassis and shields may become "live" if the mains con-nection is made the wrong way round. It is safest to consult a Hi-Fi dealer if in doubt. (Many TV rental companies will supply isolating transformers by arrangement.)

by making special connections inside the set, so as to take an audio feed from a point ahead of the output stage. In some cases, the signal diode load resistance can be "tapped down," so as to reduce non-linear distortion by improving the diode operating conditions at deep AM modulation excursions. Note that safety precautions are necessary with mains-operated sets. Isolating transformers may be required (ex RS components etc).

We may summarize by saying that the sideband cutting on an AM set is due largely to the sharply tuned IF (intermediate frequency) filters which have to be introduced so as to make the set selective enough to separate adjacent station channels at

8 kHz intervals. In the service area of a powerful local station during daylight, when distant stations are usually not received owing to the absence of the upper-atmospheric reflecting layer which forms after dark, higher-quality AM sets can be used in which the selectivity is reduced. This enables the more extreme sidebands to be received, with consequent improvement in the high-frequency sound reproduction.

Television sound Bands I and III are AM in Great Britain, but there is no need for the extreme selectivity required on the normal AM medium-wave broadcasting bands, and the IFs can be tuned fairly broadly so as to give a reasonably good high-frequency response. It is necessary, however, to see that the "skirt" of the response on the side adjacent to the vision channel is not so wide as to allow picture-line and frame-synchronizing pulses to break through.

Not all TV sets may be suitable for modification to give a high-quality sound output in this way. A further difficulty may be that the amount of mains HT supply smoothing provided may be adequate for reproduction on the small loudspeaker in a table TV set cabinet, but the fuller bass response afforded by a subsequently added high-quality amplifier and loudspeaker may give intolerable hum, unless the TV set power supply smoothing is considerably improved. Many better-class TV sets can be rearranged and/or modified to give a really good high-fidelity sound feed to a separate amplifier and loudspeaker system. As we have mentioned, the sound quality obtainable from modern TV studio sound equipment and transmitters is potentially very good and specialized UHF sound tuners may be used to feed Hi-Fi amplifiers. Tuners are made by Motion Electronics Ltd or Rola Celestion "Telefi." The latter needs no aerial, as it picks up TV sound by magnetic coupling only from the TV set IF stage.

Stereophonic sound broadcasting is carried out economically on FM by using a sub-carrier on the normal single FM transmitter. AM stereo experiments have usually demanded two separate transmitters and are not regarded as an economic proposition. There have been other suggestions for AM stereo transmissions but these do not seem to have progressed beyond the experimental stage. Regular stereo broadcasts are now put out by a number of stations in Europe and the USA, using the FM sub-carrier technique on single transmitters. Comparatively simple additional decoder circuits can be added to the discriminator circuits of many FM receivers in order to produce two high-quality stereo channels, but the best decoders involve fairly elaborate circuits.

The main carrier system of an FM transmission will give a good single-channel (monophonic) programme on a normal set, where the stereo result is not required. The system is thus termed "compatible."

We have noted that it is unfortunately true that, except for listeners residing in interference-free areas close to transmitters, fairly advanced radio-receiver circuits and good outdoor aerials are required if high-quality reception is desired. Normal AM broadcast receivers have long incorporated features such as band-pass IF tuning filters, image-suppression circuits, automatic volume control (to combat fading), anti-drift oscillators etc. A good modern FM receiver should have many other features, such as precisely designed band-pass tuning filters of adequate band-width, an accurately stabilized oscillator and/or automatic frequency control, proper limiting action, adequate capture-ratio and a carefully designed discriminator; preferably also including inter-station muting or "squelch" reduction.

A good set, however, cannot give its proper designed stereo or other performance when fed from an inadequate aerial system, particularly in areas where the field strength is low owing to distance or screening effects etc.

It is important that the buyer of high-fidelity equipment should be able to assess the merits of the various types of commercial receiving equipment on the market. Examples of well-designed aerial systems and receivers are therefore described in the following pages.

Aerial systems

A good broadcast band AM aerial system basically consists of an adequate vertical length of wire extending outdoors to a reasonable height above ground. Any horizontal extension of the vertical wire from its top end adds capacity and thus helps to complete the circuit back to earth, but is not usually a critical part of the aerial design. Many good broadcast (long, medium and short waveband) aerials thus consist of a single 10–15 ft vertical rod or wand erected as high as possible on top of a building or on a mast. The down-lead through a building to the receiving set does not contribute much to signal pick-up and may, in fact, pick up locally generated electrical interference within the building. The

Belling-Lee "antiference" broadcast band aerial was designed as a vertical wand of the above type with a screened down-lead. Impedance matching transformers were used at top and bottom of the down-lead. These transform the relatively high aerial impedance to a lower value, so that the relatively high-capacity down-lead does not cause any appreciable shunt loss. The impedance is stepped up again at the set terminals to a value which suits the normal receiver's input circuits. A coil on a ferrite rod a few inches long can be used to replace a longer wire, particularly in areas of high field strength.

It must be noted that broadcast radio waves are transmitted by radiating wires or elements that produce electromagnetic radiations which are "polarized" either in the vertical or in the horizontal plane, according to the channel requirements allocated by the ruling authorities. This plane of polarization means that a vertically polarized wave will theoretically give the maximum induced signal in a vertically disposed aerial conductor and, similarly, a horizontally polarized wave will give the maximum signal in a horizontal conductor. Most broadcast band aerials are in the form of large vertical masts which radiate vertically polarized waves. There is some evidence that electrical interference often tends to have more vertically polarized power than horizontal and thus horizontally polarized transmitting arrays and receiving aerials give a slightly better signal-to-interference ratio than vertical aerial systems. The question of polarization and orientation of aerials becomes much more critical on TV and FM VHF and UHF transmissions. Mutual interference between adjacent TV stations can sometimes be minimized by adopting vertical polarization for one station and horizontal polarization for a neighbouring station. Owing to the "capture-effect" on FM, interstation interference between adjacent FM service areas is not usually a problem and horizontal polarization is almost universal for FM broadcasting. The VHF and UHF aerials differ from the long-, medium- and short-wave broadcasting single-wire aerials described, in that basically they consist of symmetrical "dipole" rods and precisely spaced adjacent reflector and/or director rods. The feed connection to the set is taken from the adjacent centre connections of the two dipole rods. The simple dipole is "bidirectional" in one plane, with a "figure-of-eight" polar pick-up pattern. The addition of front director and rear reflector rods spaced correctly from the signal dipole make the aerial more sharply unidirectional, according to the number and spacing of the subsidiary rods. Dipoles are fairly sharply tuned "wave-resonant" arrays, which give a fair measure of selectivity to the transmission for which they are designed. The length of the rods determines the resonant frequency and the diameter controls the bandwidth. A structurally strong aluminium alloy rod of about $\frac{3}{8}$ in. diameter usually gives an adequate TV bandwidth. The overall length of a dipole is given by the following formula—

$$L = \lambda \times 1{\cdot}57 = \frac{474}{f} \text{ ft approx}$$

where λ = radio wavelength in metres
  f = radio frequency in megahertz.

Figs 6a and 6b give dipole reflector and director details for British FM broadcast channels, and show one type of combined AM/FM array.

Thus if more than one channel is to be received by tuning the receiver, either different arrays or a composite multiple array must be used. The array must also point towards the desired station and, particularly in the USA, elaborate remotely controlled rotatable aerial systems are used. Mast-head transistorized RF amplifiers can also be used in order to boost the signal from the aerial before it traverses the down-lead, in which a certain loss of signal strength inevitably occurs and which may pick up some electrical interference. The down-lead cable from a dipole is of either the 75-ohm screened coaxial or the screened low-capacity twin-lead type. In either case the spacing and insulation of the conductor(s) and/or shield is critical and must be correctly designed. The input circuit of the receiving set must also be correctly designed to match the cable impedance, and the cable must not be "tapped" or looped on to provide another outlet without the use of special matching coupling units.

We see from the above considerations that, for most high-quality enthusiasts, there is no substitute for a well-designed and properly sited aerial system. The results which can be attained and the type of aerial required in any given district can vary considerably, particularly for VHF and UHF transmissions. Advice from local radio amateurs and Hi-Fi dealers may be very helpful in this respect.

In general it pays to buy a good-grade commercial aerial and

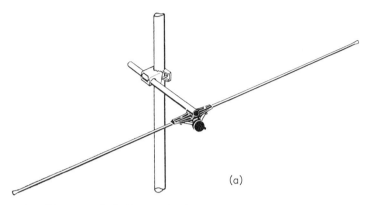

(a)

Fig. 6(a). A simple half-wave dipole VHF FM aerial for use in areas of moderate to high signal strength. The FM band II extends from 87·5 to 100 megahertz and the total width over the rods is about 5 ft. Standard low-loss 75-ohm coaxial cable is normally used but the equivalent shielded twin may also be used in some cases.

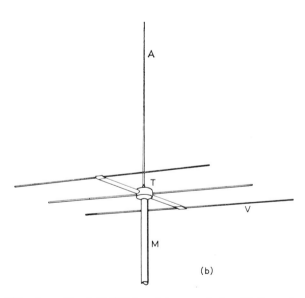

(b)

Fig 6(b). A combined AM/FM aerial array. A is a 10 ft conducting fibreglass wand AM rod coupled to the 75-ohm coaxial cable by a transformer T. V is a three-element FM array also coupled to the cable.

have it properly erected, using, needless to say, the correct specified down-lead and plug connector at the set input terminal. Aerial manufacturers will often provide instructions and technical advice in cases of difficulty.

The question of the risk of damage by lightning may be considered, particularly in very exposed sites. Modern aerials are not unduly susceptible to lightning strikes, but some small risk still

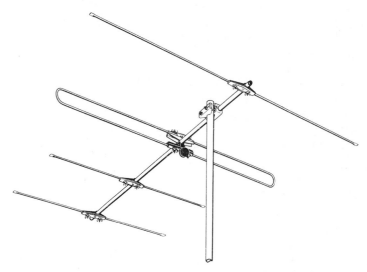

Fig 7. An efficient folded dipole with a reflector and directors, giving a suitably directional response for areas of lower signal strength or where excessive interference or multi-path propagation problems exist.

remains. It is thought that transistor mast amplifiers may be damaged by comparatively small lightning discharges which may not cause any other noticeable damage. It is often possible to insure aerial systems against lightning damage for a very small premium, or the risk may be covered in comprehensive domestic insurance policies issued in some cases. The possibility of damage by lightning discharges has caused anxiety in some districts as to the likelihood of damage to input transistors and masthead transistor booster amplifiers. In practice it seems that the isolating effect of input transformers and the series inductance of leads is sufficient to avoid such damage except in particularly unfortunate cases.

Tuner principles

The circuit line-up of a good FM tuner might be along the following lines—

1 The 75-ohm coaxial or balanced twin low-loss feeder from the aerial array is accurately matched to low-noise input stage, and further tuned band-pass stages provide adequate selectivity ahead of the mixer/oscillator stage, so as to provide adequate rejection of unwanted incoming signals, particularly those of the IF frequency (10·7 MHz) and the IF image frequencies corresponding to the signal being tuned in. The oscillator is normally tuned 10·7 MHz above the incoming signal. The image frequency will be 10·7 MHz above or below the oscillator and thus can pass into the IF amplifier unless attenuated.

2 The stability and accuracy of the oscillator are important, whether the circuit is of the fixed-tuned channel or the continuously variable tuning type. Oscillator temperature drift may be compensated by the adoption of negative temperature coefficient capacitors, by crystal control, by the use of a quarter-wavelength tuned line or "trough," or by the use of automatic frequency control (AFC) provided by a reactance tube or diode varactors whose characteristics are varied by the discriminator output in such a manner as to bring the oscillator accurately back on tune in the event of any oscillator drift occurring. Tuning of the RF and oscillator stages is accomplished by moving low-loss sintered ferrite cores axially inside the tuning coils, ie permeability tuning, by ganged tuning capacitors or by varying the bias on varactor tuning devices.

3 Two or three IF stages coupled by band-pass filters centred on 10·7 MHz follow the mixer stage. The first stage may have variable μ characteristics so as to enable a measure of automatic gain control to be applied. The last IF stage should be arranged to saturate sharply above a certain threshold signal so as to give an effective limiting action. A good limiter stage is a very important attribute of an efficient FM tuner, as all traces of unwanted AM signals, including most types of interference must be eliminated.

4 A carefully designed discriminator circuit of the Foster-Seeley phase-discrimination type or the ratio-detector type will produce an undistorted audio signal from the constant-amplitude frequency-modulated signal fed to it from the limiter.

The carrier frequency components are eliminated from the output, but a DC signal is produced which drops to zero when the receiver is exactly on tune. This DC signal has a magnitude proportional to the amount of mistune and the polarity depends on whether the receiver is tuned above or below the correct frequency. This DC signal from the discriminator can be extracted and used to operate a tuning indicator, as well as controlling an AFC reactance device. The polarity of the DC signal is used to operate some excellent forms of precise tuning indicator. Normal tuning indicators may only be a DC meter indicating a minimum "dip" for exact tuning. These do not indicate the direction in which the tuning control needs adjusting in order to restore or achieve correct tuning and it is necessary to de-tune in order to check that a minimum dip is indicated.

The AF response is given a "pre-emphasized" rising characteristic at an FM transmitter in order to improve further the signal-to-noise ratio, as the high noise frequencies can be attentuated by an equivalent amount after the receiver discriminator. These simple RC series-equalizing combinations are specified most easily in terms of their time constant, ie the time in microseconds for a given charge on a capacitor to fall to I/N times the initial value (N is the natural logarithm base value 2·71 . . .). European and British FM transmissions use a 50-microsecond (μs) pre-emphasis. This fact should be remembered when considering imported sets.

Some designers used to prefer to use vacuum tubes (valves) throughout FM tuners, some used a mixture of tubes and semi-conductors, but the majority now use semi-conductor (solid-state) devices exclusively. Good designs may be achieved in each case and it is difficult to do other than to judge a product on its merits.

5 Inter-signal quieting or squelch circuits are used as it is desirable to prevent or inhibit the action of AVC for inputs below the usable minimum, so as to avoid an up-rush of noise when tuning between signals. The "squelch" circuits may provide an amplitude threshold or "delay" (so called) in the AVC voltage-generating stage which, in effect, provides an input "gating" effect which prevents the receiver coming into action unless the input signal exceeds a certain minimum value. This value ideally should represent the point at which limiting commences.

Typical high-quality FM tuner quantitative specification figures
Frequency range: 88 MHz to 108 MHz (Band II).
Sensitivity: 2 microvolts for 100 millivolts output and 20 dB quieting.

6 microvolts for full quieting (40 dB).

Audio response: Flat to ± 1 dB from 30 Hz to 20 kHz (allowing for the standard 50 microsecond de-emphasis characteristic).

Limiting characteristic: limiting commences at inputs above 2 microvolts.

Constant output level produced for inputs above 3 microvolts.

AFC characteristics: signal held constant to ± 1 dB for tuning centre frequency changes of ± 120 kHz approximately.

Image and IF rejection: 40 dB and 80 dB respectively.

Capture ratio: 3 dB.

AM suppression: 35 dB.

Stereo-channel separation: better than 35 dB at 1 kHz.

Fig 8 shows a high-quality FM tuner which should exceed this specification and will give the best results under adverse conditions. Local conditions and the number of FM signals receivable to some extent determine the elaboration of the tuner which you will wish to use. Again, your Hi-Fi dealer can often advise you on the most suitable tuner.

FM stereo multiplex receiver units
We have mentioned previously that the standard method of stereo broadcasting now used for radiating programmes in Britain, the USA and several European countries, notably France, Germany and Holland, is the FM "pilot tone" system in which the original two stereo sound channels (A and B) are subjected to a multiplexing or encoding process by which the instantaneous sum ($A + B$) is in effect transmitted by frequency modulation of the main carrier, whilst the instantaneous difference ($A - B$) is transmitted by a sub-carrier, in addition to a super-sonic pilot tone of 19 kHz. Standard monophonic FM receivers produce the sum ($A + B$) of the left- and right-hand stereo

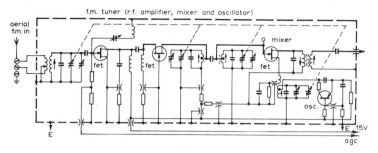

Fig 8. *The schematic of an advanced modern stereo FM tuner* (continued on the following pages). *This includes a number of advanced design features which may be essential to obtain the best results in areas of moderate or low field strength. Simpler tuners may be used where the field strength is high, but there is a trend towards the use of sophisticated tuners of this type when the best stereo results are desired. The RF and mixer stages use three low-noise field-effect transistors. Tuning is by a ganged variable capacitor.*

channels at the discriminator output without any circuit modifications being needed. If the stereo microphone and mixing technique is skilfully arranged to this end, the sum signal can usually be made to give an acceptable high-quality monophonic sound signal. There may, however, be cases in which the need to produce the best possible stereo sound "picture" may mean some sacrifice of the monophonic sum signal acoustic "balance." This may possibly be justified on the grounds that listeners who want the best reproduction will be prepared to invest in stereo equipment and that the need to produce the best possible stereo result must be the paramount requirement. Generally, a good FM receiver will have sufficient band-width to accommodate the pilot tone and the difference channel in addition to the normal main carrier. An add-on decoder unit can usually be fitted on to a suitably designed normal discriminator circuit in order to extract the two original left- and right-hand stereo *A* or *B* signals. Most of the leading manufacturers of tuners allow chassis space or otherwise make provision for the easy addition of a stereo decoder unit which they have available. Decoder circuits have also been published by manufacturers and technical journals.

Fig 8 gives a tuner circuit using advanced modern design concepts. Many such designs include an electronically operated indicator which shows when a stereo programme is being received from any given transmitter. The non-linear distortion produced

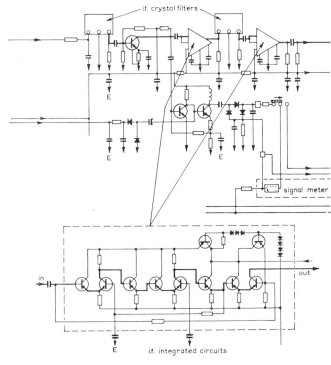

Fig 8 (contd.). *The IF stages use crystal band pass filters which give a stable wide-range response with high rejection of unwanted signals. Good limiting characteristics are obtained. An advanced discriminator and inter-station noise-reducing squelch circuits are used.*

by the encoding and decoding process is claimed to be substantially as good as that from the best monophonic broadcasts, and the stereo channel separation is usually between 35 and 45 dB. This is slightly worse than is achieved by the best stereo tape-recordings, but is substantially better than that of most stereo gramophone pick-ups.

From the broadcaster's viewpoint, the addition of the sub-carrier does not normally represent a very difficult modification of the transmitter, but some of the available power is diverted into the sub-carrier and thus the effective service area of the transmitter may be reduced. This is important if a national network is being planned, as extra transmitters may be needed to maintain the same coverage. The main complication is undoubtedly in the

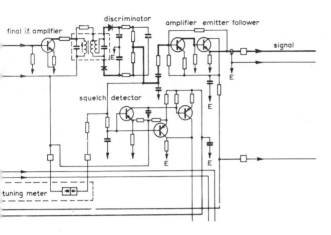

radically different and much more demanding microphone studio and line transmission techniques which good stereo programmes demand. Stereo tapes and records do, however, provide a convenient source of programme if a sufficiently large library of recordings is available.

The two decoder outputs are, of course, fed to two separate stereo amplifiers and loudspeakers, which must be balanced in the usual way. The advantage of a panel indicator which shows when a stereo signal is tuned in is that the decoder and amplifier channels can then be switched to the stereo position. Otherwise, one may continue to listen to the $(A + B)$ monophonic version of the stereo broadcast. This may be desirable in marginal cases where noise is too severe for the full stereo result to be enjoyed.

Special characteristics of frequency modulation

We have seen that one of the great advantages of FM is that the effects of interference are greatly reduced. It is fairly easy to see that normal forms of AM electrical noise and static interference are greatly reduced by FM peak limiting. It is not so easy to see how the effects of adjacent channel station interference are also greatly reduced by FM, because of the so-called "capture effect," so that a slightly stronger signal can "capture" the receiver and reduce the weaker competing signal to complete inaudibility, though it may still cause some distortion of the "winning signal."

With AM broadcasting reception, an adjacent channel signal of 1/100 of the strength (-40 dB) of the desired signal can still

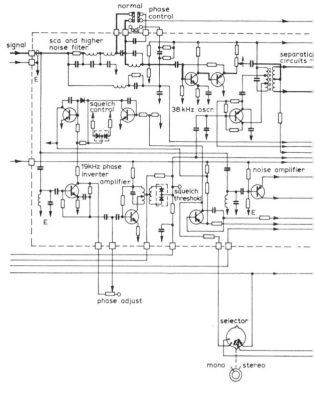

Fig 8 (contd.). *The decoder circuits produce L and R channel signals with low distortion and cross-talk levels. The sub-carrier frequency and harmonics are effectively suppressed from the output by suitable filter circuits.* (Contd. on facing page.)

give rise to an annoying inter-carrier beat frequency whistle, whereas with FM, a ratio of 2/1 or less in relative signal strength can be sufficient for full capture by the stronger signal. If two transmissions pass through the RF, mixer and IF stages of a FM set, as they will if they are of nearly the same frequency, the resultant RF wave will generally be both frequency- and amplitude-modulated. A good limiter will remove the amplitude modulation, but the phase modulation remains. However, it can be shown that the resultant vector becomes, in effect, frequency-modulated at the frequency of the stronger of the two incoming transmissions, even though the difference in amplitude is quite small. Thus the receiver locks on to the stronger signal, which is said to capture

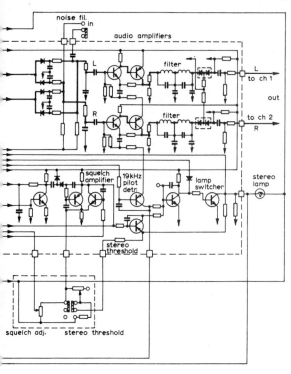

the set. Owing to the selectivity of the RF and other stages and
the relatively wide-frequency channel-spaced allocations normally
given for stations within normal mutual range, it is likely that adjacent transmissions will produce much greater ratios of strength
than 2/1 when a particular station is tuned in. However, the effectiveness of the capture action and the normally fairly short range
of VHF transmissions mean that a number of stations can be
operated at the same frequency provided that they are out of
mutual range. Freak reflections from the ionosphere do, however,
sometimes bring a distant transmission in at sufficient strength
momentarily to capture a normally much stronger signal.

FM stereo decoder circuits and specifications

The unit may be electronically or mechanically switched on to the
output of the discriminator without the de-emphasis circuits.
A high-input impedance circuit is provided so as not to load the
discriminator. In some cases the emitter-follower output from

this stage drives a quadruple diode ring modulator via a low-pass filter designed to remove the 19 kHz sub-carrier frequency. The circuit is also designed to provide overall phase compensation of the signal. The diode ring modulator is "switched" by the output of a synchronized 38 kHz oscillator. A tuned circuit in the input stage feeds the 19 kHz received pilot tone via an amplifier and rectifier in order automatically to bias the diode signal switches to the "on" state for stereo. The 38 kHz pulses produced synchronize the oscillator switching signal correctly with the pilot tone. The left and right channels are recovered directly, the correct balance between sum and difference signals being automatically restored. Output filters prevent any sub-carrier leak to amplifiers. Other decoder circuits are also employed, including phase-locked-loop ICs (see page 41).

Typical outline specification of decoder performance—

De-emphasis: 50 μs (European).
            25 or 75 μs (N. America).

Output: 100 microvolts at 30 per cent maximum excursion.

Stereo cross-talk: better than 30 dB at 1 kHz between channels.

Pilot tone suppression: at least 30 to 40 dB suppression of 19 kHz pilot tone and 38 kHz switching tone in output. This is important as harmonics may cause whistles with tape-recorder bias frequencies.

High-quality AM tuner design

We have noted that, because of the close channel spacing, the general overcrowding and the large distances at which stations can be received, it is necessary for an AM radio receiver to be provided with a very high degree of selectivity when conditions make this necessary. This, unfortunately, leads to a severe loss of high frequencies because of the cutting of the outer side bands. When conditions are good and no severe inter-station interference is present, the band-width may be increased. Thus variable selectivity is highly desirable. This can usually be satisfactorily obtained by switching the IF band-pass filters, using tertiary or auxiliary windings designed to widen the pass-band.

   AM reception is also much more open to noise interference and no effective form of noise-limiter can be adopted, unlike FM where very effective noise-limiting circuits can be used in the receiver. A switchable sharply tuned "dip" circuit tuned to around 8 kHz is also desirable to remove inter-channel "beat" whistle on AM.

For listeners residing at a distance from the desired stations and in areas where an appreciable amount of man-made electrical interference exists, a good high AM aerial rod is a valuable initial investment. This and other forms of aerial system have been dealt with in the section on aerials. The input circuits of an AM superhet receiver must have a good degree of selectivity, particularly at the IF frequency (470 kHz), as anything received at this frequency will otherwise pass straight into the IF amplifier. The "image" frequencies at ± 470 kHz from the oscillator frequency must also be attenuated.

The general operating principles of the superhet receiver and the detailed design of the various stages, filters etc is dealt with in comprehensive books and articles to which the reader may be referred.[4,6] The input-tuned circuits are usually combined with a low-noise RF input stage which is tuned by a ganged variable capacitor or a variable permeability system in synchronism with the oscillator, whose circuits have, of course, to be tuned exactly 470 kHz above or below the frequency of the incoming signal, in order to transfer the audio modulation to a carrier having the 470 kHz IF frequency. The production of this IF "beat" or difference frequency depends on the non-linear RF transfer characteristic of the mixer stage. This is a further reason for the use of a good RF stage with adequate selectivity, as this attenuates unwanted signals ahead of the mixer and reduces the chance of cross-modulation of wanted and unwanted signals in the mixer. In addition, the mixing process is more efficiently carried out and a better noise ratio is achieved if the wanted signal is raised to a suitable level by the RF amplifier.

The RF mixer and IF stages are used to give automatic gain control (AGC) by the use of variable-μ valves or alternative semi-conductor circuits. AGC is essential to prevent violent changes in the reproduced level due to fading and differing station strengths when tuning. AGC is generally derived at the detector stage by the use of a second diode with suitable RF decoupling stages. The impedance loading placed on the signal diode in an AM detector circuit is extremely critical if severe non-linear distortion of heavily modulated signals is to be avoided. The important criterion is the ratio of the AC and DC impedance

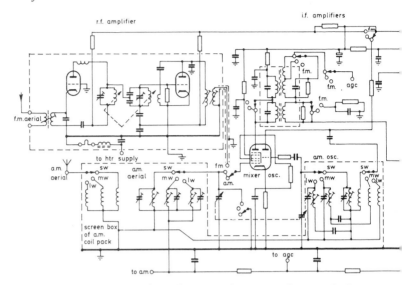

Fig 9. *A modern high-quality valve AM and FM tuner functional schematic. Points to note are high selectivity in an RF stage ahead of the mixer, variable band-width IF transformers with tertiary windings and low-distortion diode detection system. These features, respectively, eliminate input image frequencies, whistles etc, reduce sideband cutting on strong local signals and enable heavily modulated loud signals to be demodulated without distortion.* (Contd. on facing page.)

loading which follows the signal diode. Thus the value of the coupling components between the diode and the first AF stage and the input loading due to an AGC diode and its circuits may have to be carefully controlled. The effect of tuning indicator circuits may also have to be considered if these are effectively coupled at this point. A most effective way to achieve a high AC/DC ratio of diode load is to "tap down" the load by taking the AF signals output from a small fraction of the resistive load (25 per cent or less), thus "diluting" the effect on the diode of the AC load due to coupling capacitors etc. Push-pull full-wave diode detector circuits are also used.

AF noise suppressors

Various attempts have been made to eliminate or reduce impulsive noises which get through to the AF circuits. Generally these take the form of a preset peak-clipper designed to prevent the crests of large interference "spikes" being reproduced. Another system

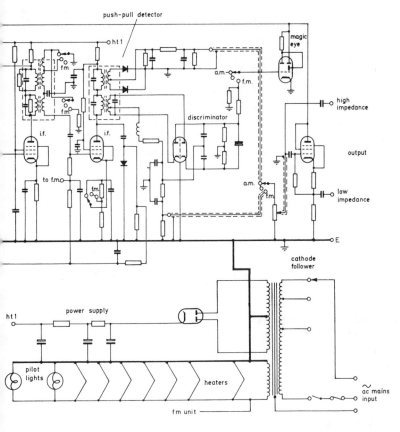

is a "rate of rise" clipper designed to come into operation for a millisecond or so to cut out a "spike" with a very sharp wavefront. It is difficult to make either of these systems effective without affecting the quality of reproduction of music containing transients of the same nature as impulsive interference. The use of these circuits has thus largely been restricted to communications and short-wave receivers where high-quality reproduction is a secondary consideration.

Specification of a typical high-quality AM tuner

Tuning range: Long wave 2070–800 metres.
Medium waves 588–185 metres.
Short waves 1 2·2–6·6 MHz.
2 5·8–18·5 MHz.

AF response (*narrow band*): 5 kHz bridged T low-pass filter with 18 dB/octave minimum cut-off above 5 kHz. Wide band, minimum loss up to over 12 kHz (6 kHz and 12 kHz 3 dB points respectively).

Dip circuit tuned to eliminate 8 kHz with an attenuation of at least 20 dB. Loss at 7 kHz preferably not to be more than 6 dB. Output level 100 millivolts at 30 per cent modulation into 50 k-ohms (approximately).

The tuning indicator may be automatically switched out for wide-band response, as it will not normally indicate the centre point of a wide-band tuner and may thus be misleading. Delayed AGC is provided. The delay means that an initial offset or gating characteristic is provided so that noise and signals below a certain level do not operate the AGC.

Sensitivity: Better than 20 microvolts for 10 dB signal to noise ratio.

Better than 100 microvolts for 40 dB signal to noise ratio.

IC broadcast stereo decoders

Integrated circuits are now readily available which greatly simplify the home construction of two-channel stereo decoders. These circuits can be added on to the discriminator output on many types of FM tuner, where these are giving good noise-free monophonic reception from stereo broadcasts. It must be borne in mind that the noise level on stereo will be raised considerably, due to switching noise and the extra band-width required. Fig 10 shows the application of a new type of IC decoder made by Motorola and also National Semiconductor which only requires the addition of some non-critical resistors and capacitors to make a decoder which gives a very good performance, based on the phase-locked loop principle. This provides, in effect, a high "Q" servo amplifier, using the 19 kHz pilot tone to produce the 38 kHz phase-locked sub-carrier to drive the decoding transistors. The only setting-up required on first receiving a stereo signal is to adjust the pre-set 5 k-ohm resistor until decoding takes place, no further adjustment normally being required. A noisy stereo programme can be converted to mono by closing the "mono" switch, and a stereo "beacon" lamp of filament or solid-state type automatically lights when decoding is taking place. This integrated circuit contains over 50 transistors as well as internal resistors and capacitors. It is a remarkable example of the compact and economical "circuit building block" ICs which now perform complex functions. These circuits with printed mounting boards and the extra passive components required are available from several suppliers (eg Integrex, Derby; Jermyn Industries, Sevenoaks).

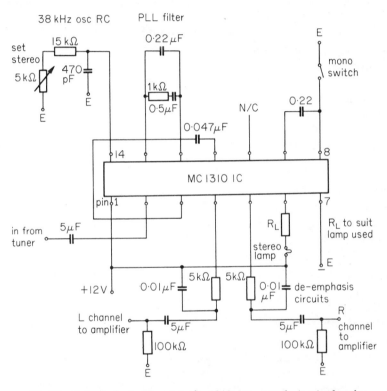

Fig 10. *A typical circuit using the* 1310 *integrated circuit decoder for use with suitable FM tuners in order to produce two stereo broadcast programme channels. It uses the phase-locked loop principle to regenerate and to synchronize the* 38 *kHz sub-carrier* (Motorola Semiconductors Ltd).

One point to be noted is that any de-emphasis circuits following the discriminator must be removed or the pilot tone and stereo modulation will be seriously attenuated. De-emphasis is provided in the stereo decoder circuits.

Means of reducing noise in stereo radio reception

Although FM radio broadcasting is known to reduce unwanted background noise to a large degree, it is an unfortunate fact that the addition of stereo can cause a deterioration of ten times or 20 dB in the signal-to-noise voltage ratio. This is due to the diversion of some of the transmitted power into the sub-channel and pilot tone and to the increased band-width, as well as the extra devices and the switching action in the stereo decoder circuits. Thus it often happens that excessive background hiss is encountered when a station giving good monophonic results puts out a stereo transmission. The only real cures are either to improve the input signal to the set by erecting a better aerial or by changing to a receiver of higher quality which has low-noise input stages and a better limiting action.

The further loading of FM multiplex transmitters with additional encoded signals to carry quadraphonic stereo may again tend to increase the noise slightly.

Other forms of stereo noise occur due to "beat notes" between harmonics of the pilot tone in the decoder and other signals which are normally supersonic. Additional filters are required to remove these whistles or warbles, and these filters are incorporated in better-grade stereo receivers. Many simpler sets or designs and economy kits for home construction may not incorporate such filters and thus some constructional details are given in the next section.

Whistle filters for stereo decoders

Two principal sources of whistles are—

1 Adjacent channel FM signal beat notes produced with the fifth harmonic of the pilot tone generated in the decoder.

The adjacent channel-frequency spacing is 200 kHz and this will pass through the discriminator and form 10 kHz and lower audible beat notes in the decoder circuits because regeneration of the 38 kHz sub-carrier from the 19 kHz pilot tone is normally performed by a non-linear process which generates numerous harmonics of 19 kHz, the 10th harmonic being 190 kHz. Advanced tuners, such as that shown on page 34, include effective low-pass filters which cut off the frequencies above 55 kHz, which include not only the normally unwanted SCA subsidiary communication modulation used in the USA but also 200 kHz and lower carrier beat notes due to adjacent powerful FM stations which manage to break through the receiver front-end circuits at a level which, although low, is still sufficient to cause annoying audible whistles and warbles or "birdies" in the manner described above.

2 Leaks of pilot tone, sub-carrier and other frequencies which may appear in the tuner output channels and form further audible beat whistles with the bias frequency of a tape-recorder when this is used. Again, ambitious tuners use low-pass filters in the audio output channels which are tuned to cut off all frequencies above about 17 kHz. Quite good-quality multi-stage low-pass filters or piezoelectric ceramic filters must be used here in order to give 30 to 40 dB attenuation of 19 kHz and higher frequencies without reducing the frequencies between 10 and 17 kHz which are so important to high-quality reproduction.

It is possible to purchase ready-made filter modules for addition to existing tuners (eg Toko UK Ltd, Slough, Bucks, and Messrs Harrogate Radio, Harrogate, HG1 2DB).

However, now that ferrite pot cores may be obtained from many manufacturers and from stockists (such as Henry's, Edgware Road, London) many home-constructors may wish to construct their own filters. The pot cores may be adjustable types and the spools are easily wound by hand to make up filters with standard capacitors,

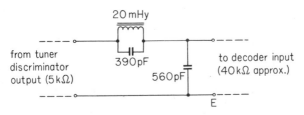

Fig 11. *FM inter-channel whistle filter. This low-pass filter removes beat frequency whistles and warbles on stereo. The attenuation is 0 dB up to 25 kHz, −6 dB at 40 kHz, −30 dB at 55 kHz and over. Approximately 250 turns on adjustable ferrite pot core (Henry's Radio etc). Tune core for maximum reduction of whistles without loss of stereo pilot tone. Reduces unwanted broadcast pick-up. 1000 pF capacitors wired across base-emitter on pre-amp input transistors also help.*

the units being fitted into standard metal screening boxes, which are readily available from stockists. Figs 11 and 12 give details of the various stereo filters and attenuation figures obtainable.

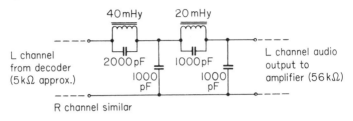

Fig 12. Bias whistle filter for tape-recordings from FM stereo. This filter is a low-pass and band-elimination type which may be inserted in each audio channel output from an FM stereo decoder in order to eliminate beat whistles caused by 19 and 38 kHz pilot tone and sub-carrier leaks from the decoder beating with tape-recorder bias oscillator frequency (100 kHz etc). Wind ferrite pot cores with 350 and 250 turns respectively. Tune for maximum whistle elimination, without loss of higher audio frequencies (Henry's Radio etc). Attenuation is 0 dB at 10 kHz, −1 dB at 15 kHz, −4 dB at 17 kHz, −22 dB at 19 kHz, −35 dB at 38 kHz. One filter per channel. Fit in separate metal shielding boxes.

Quadraphonic broadcasting

It is now considered to be technically possible to broadcast four discrete channels by means of an extension of the FM multiplex base-band and sub-carrier system used at present for two-channel stereo broadcasts (see Fig 13). However, in order to achieve compatibility with existing systems, as well as present stereo and mono receivers, an SQ matrixing encoding system has been tried

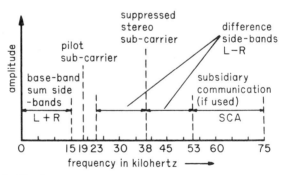

Fig 13. The frequency spectrum transmitted by normal two-channel FM stereo broadcasters. The upper SCA channel is used in USA for broadcasting commercial "background music", etc using special decoders. Various signals may cause whistles in domestic stereo decoders unless they are filtered out.

as well as various FM multiplex 4-channel systems. Typically, the coded 4-channel signals are fitted into the FM transmitted spectrum, using a 76 kHz additional sub-carrier modulated to occupy the top frequency space previously occupied by the SCA subsidiary channel information. It is thought that these transmissions fit into the present 200 kHz FM station frequency spacing and that the increase in noise due to a reduced signal-to-noise ratio as compared to present two-channel stereo broadcasts would be not more than 6 dB, an amount which could be recovered by an increase in transmitter power and by improved aerials and receivers. Provided that acceptable compatibility is achieved, it seems likely that quadraphonic broadcasting will become widespread, but probably not for several years.

FM inter-channel squelch circuits

The users of tuners without this facility are made very aware of the up-rush of noise as one tunes between stations of normal strength. This is due to the loss of the normal receiver limiting action which occurs when a strong radio signal is applied to the IF limiting stages. This is equivalent to a strong AVC action, turning down the gain effectively for a strong signal. Some highly sensitive receivers may also have additional AVC circuits or local station attenuating circuits which avoid the over-loading of the receiver input stages.

This disconcertingly loud inter-station noise is suppressed by squelch circuits in the following manner: The IF signal level is detected at the input to the limiter stage. It is rectified, amplified, inverted and made to switch a transistor across the audio output from the discriminator when the received radio signal falls below a predetermined low level. The effect is that the set is muted until a signal is tuned in which is strong enough to initiate the limiting action and to give an acceptably noise-free programme. At this level, the muting or squelch transistor is switched off and no longer short-circuits the audio output which feeds the decoder, in the case of stereo tuners. A number of circuits have been designed to perform squelch functions and integrated circuit packages are available (eg Lithic, Europartners, Slough, Bucks).

3

Gramophone records and equipment

The popularity of gramophone records as a source of home entertainment has always been considerable and the advent of long-playing microgroove records pressed from low-noise unbreakable vinyl plastic gave them a further sharp fillip. The gramophone record system has been the subject of very extensive development and at their best the results achieved are hard to rival by any other recording medium. Gramophone records are also very convenient to handle and store and they permit of readily locating any part of a recording by the use of a comparatively simple micrometer type of groove-locator operating on the pick-up arm.

Automatic record-players can now provide very high-quality reproduction and can be loaded up with long-playing records to give hours of entertainment of durations in excess of that obtainable from similar-quality tape-players. These, of course, may involve tedious rewinding, spooling and tape-threading procedures except for cassettes or cartridges.

Gramophone records now, of course, provide 2/4-channel stereo by modulating each groove wall independently and by employing a dual type of pick-up with mutually perpendicular sensing elements actuated by a single stylus. Such systems can now be made to work very well, but the magnetic tape system has an advantage over the gramophone record for stereo or multi-channel recording as regards the number of channels and their mutual separation.

Recording and reproducing systems place very stringent requirements on the mechanical transport system which moves the recording medium past the transducers and a very high order of long- and short-term speed accuracy is essential. Long-term speed inaccuracy means that playback and recording speeds may differ. This can introduce errors of musical pitch and also alters the duration of a programme, which is sometimes important. Short-term variations appear as "wow" if they have a sub-audible cyclical repetition rate, or as "flutter" if the cyclical rate is in the audible band. As an indication of the stringency of the requirements, it may be noted that a frequency variation of 0·05 per cent wow may be noticeable at a critical part of the frequency band, such as 2 kHz.

To achieve this speed accuracy means not only that mechanical parts must be very accurately made to size but also that friction and vibration must be very low. This implies near-perfect shafts, bearings guides and pivots in all parts of the motors, drive and transport mechanisms generally. It is possible to achieve this sort of mechanical perfection by meticulous mechanical fitting and the use of very high-grade machine tools and gauging equipment. For quantity production of gramophone turntables and tape-decks, individual fitting of this type obviously is not possible, and carefully planned high-grade machining and testing equipment have to be installed.

From a user's point of view, it is obviously necessary to treat mechanisms with great care so as to avoid any possibility of destroying the accuracy by damaging surfaces, bending turntables or shafts etc. In particular, rubber or synthetic-rubber-covered drive wheels must not be left in engagement with shafts etc for long periods at rest in case they take a permanent "set," with disastrous results on the accuracy of their driving properties.

It is also very important to mount motor- or tape-decks in the manner recommended by the makers, using the exact fixings supplied, so as to avoid any risk of distorting frames or transmitting external shocks or vibrations to the turntable or deck. In some cases, the chassis or mounting plate is floated on very flexible springs for normal use. These are by-passed by rigid fixings or "shipping bolts" when the equipment is to be moved, so as to avoid damage to the springs or equipment when in transit.

Most modern units are designed to use permanently-oil-retaining bearings, PTFE low-friction bushes and other types of low-friction pivots, and thus may not need attention. If any lubrication is called for, it should be carried out strictly according to the makers' recommendations. The old adage of "just enough in the right place" should be observed. An excess of oil inevitably collects dust, thereby violating the other important precept, which is to keep all parts of the mechanism as clean as possible and free from all the dirt, dust and fluff etc which accumulate and may adhere to both fixed and moving parts.

We will now consider gramophone reproducing turntables and pick-ups in greater detail. Tape-decks will be dealt with later (page 82).

Design and use of turntable and pick-up units

MOUNTING OF TURNTABLES

The mounting and aesthetic integration of gramophone turntables will be dealt with in the section dealing with the arrangement of cabinets and complete systems. There are a number of general points to note. Firstly, the turntable should be accommodated so as to give easy and convenient access for changing records. Thus it should be at a convenient height and, if it is in a dark place, artificial light should be provided. Good frontal access is desirable via well-fitting doors without the need to slide or roll the turntable mounting board forward. In designing the cabinet a basic decision may have to be taken as to whether standing or seated operation is desired.

It is important to isolate the turntable from mechanical shocks due to foot-falls and the opening and closing of cabinet doors etc. Any appreciable transmitted shock may cause the pick-up to "jump" in the groove, with a risk of damage. If the loudspeaker is mounted in the same cabinet, feedback oscillations may occur at a low frequency through the pick-up functioning as a vibration-sensitive transducer. It used to be desirable to confine the pick-up in an acoustically damped enclosure with stout and leakproof doors when records are being played, but there is now virtually no high-frequency acoustic "needle-chatter" radiated from the record face on highly modulated grooves. A great deal of this was due to the vertical "pinch-effect" arising from the changing angle of a "Vee" groove section during the recording cycle, giving a double frequency (second harmonic) lift to the pick-up stylus. This represents distortion, and it is well established that small amounts of audible high-frequency distortion can give rise to annoyance far beyond that which might be expected from their acoustic sound level. It is still important to mount the pick-up as far as possible away from any mains transformers, smoothing chokes etc, or mains hum may be induced, particularly in the case of variable-reluctance or moving-coil pick-ups. The turntable base-plate should also be accurately level when in its final mounted operating position. Any error in level will cause side thrust in the pick-up and may induce serious groove-tracking errors and consequent distortion. Small circular spirit levels are available for motor plate mounting.

In the case of turntables used on ships, aircraft etc a carefully designed gyroscopic self-levelling base-plate may be used. If the pick-up arm is designed to use a pivot system balanced in both horizontal and vertical planes, the reproducer will be much less sensitive to levelling errors or variations.

TYPES OF TURNTABLE UNIT

These are basically divided into single-record-players and auto-record-changers, which will take up to ten or more long-playing records. Auto-changers are necessarily more complicated and costly than the equivalent single-record-players. Many LP records require changing only every fifteen minutes and, in the case of all but some very specialized products, auto-record-players only play one side of a pile of records, the pile having to be turned over to play the other sides. Thus, unless it is particularly desired to have fairly long periods of music from records without the trouble of record changing, a single-record-player may be the "best buy" for many amateurs today. The cheapest mass-produced turntables necessarily have limitations with regard to speed constancy, vibration and total running life before overhaul. The best professional "transcription" turntables may cost from £40 upwards but there is no doubt that a really good-quality unit is a worthwhile investment for the serious record quality-lover. However, there are now some good-quality single-record-playing turntables available at about half this price.

Good-grade auto-record-changer-players are now produced at about the same price as the best single turntables. Generally speaking, for a given quality of turntable and drive, an auto-changer-player will cost one and a half to twice as much as the equivalent single-player.

Apart from the facilities offered such as the various playing speeds, fine adjustment of speed etc, and the performance specification (which can be authenticated by independent reviewers), one of the main attributes to look for in a turntable unit is solidity. A massive well-balanced turntable acts as a fly-wheel and helps to smooth out speed fluctuations, and a rigid base-plate will not be likely to have sharp mechanical resonances. These might be excited by vibration, shocks or even pick-up tracking reaction in the case of flimsy turntable base-plates. The result may be increased background noise coloured by the base-plate resonances.

Fig 14. *A modern high-quality turntable unit and plinth assembly. A very high-precision lightweight arm gives low pivot friction and provides accurate side-thrust compensation, enabling ultra-low-weight high-compliance cartridges to be used.*

Modern turntable and record-player units from established makers give very good results, approaching those obtained from the most expensive studio transcription turntables, and thus represent excellent value for money. High-quality pick-up arms capable of doing justice to modern low-tracking-weight cartridges are also provided in many cases.

The manufacture of automatic record-changers and -players*

The automatic record-changer dates from around 1930 and was invented to save the listener from having to leave his chair every two or three minutes to change a record. Records were then played

---

* Information for this section was kindly provided by Mr E. W. Mortimer and Messrs Garrard Ltd.

at only one speed (78 rpm) and with the relatively coarse groove then used the longest uninterrupted playing time available was about five minutes. Nowadays, long-playing records have popularized the semi-automatic single-record-player, with mechanized pick-up lowering and auto-stop.

Since 1930 the art of recording and reproduction from gramophone records has changed so much that it has revolutionized the whole concept of record-changer design and production. The modern record-changer or -player is a precision instrument, and this section describes some of the intricate operations used in its manufacture to provide the means for achieving the high standard of sound reproduction which is expected today. It must be able to rotate records at any one of three speeds, $33\frac{1}{3}$, 45 or 78 rpm, though the latter may become obsolete. It must handle three record sizes, 7-in., 10-in., or 12-in., and be able to sense which diameter record is to be played so that the pick-up can lower in the correct place. It must sense when the end of a record has been reached so that the pick-up can rise and move back to allow the next record to drop and be played. It must also be able to tell when the last record has been played so that it can switch itself off. In addition it should be able to play single records manually and have facilities for rejecting the record being played. A "pause" control is also useful for cueing.

The mechanical handling of gramophone records in order to play a number of them consecutively, although the basis of the automatic record-changer, is not, however, the principal criterion of its quality, which is the fidelity of the sound given by the system.

Among the many factors adversely affecting the quality of the reproduction obtained from gramophone records, three principal ones are wow, flutter and rumble, all of which are inherent to some degree in every rotating mechanical device used for sound recording and reproduction.

Wow is the term used to describe speed variations superimposed on the reproduction below about twenty Hertz and flutter occurs above twenty Hertz. Rumble is random unwanted noise below about 500 Hertz, often extending to sub-audible frequencies. These three factors originate in the mechanism used to rotate the record, and to reduce them to a practical minimum on a quantity-production basis calls for highly specialized engineering techniques, which are worthy of mention.

The high degree of amplification used between the pick-up and speaker will reproduce the slightest bearing noise as rumble and any imperfections in the driving system appear as wow and flutter. For example, a journal or thrust bearing normally considered good engineering practice would be too noisy in a mechanism used for sound-reproducing purposes, and electric motors used to drive the mechanism and turntables have to be made to much higher standards than would be required for purposes other than for the reproduction of sound.

The first essential is to make sure that all bearings are perfectly free-running, with the minimum clearance, and run with the minimum of noise or vibration. Whenever possible, oil-impregnated sintered bronze bearings are used, as these show little signs of wear over long periods of use and rarely need lubrication. The bearings need to be specially processed to meet the standard required. The bearing holes are accurately sized and finished by rotary burnishing and the self-aligning bearings as used in the electric motors are also spherically turned on the outside to make sure that the hole is true with the bearing and shaft location. Shafts which run in these bearings are specially finished, most of the turntable spindles and all rotor shafts being hardened, tempered, ground, lapped and superfinished to one micro-inch surface finish.

The clearance between the spindle and bearing bore must be very small to avoid noise which would show itself as rumble. The maximum clearance allowable in the bearings of the electric motor is ± 0.0001 in. and every motor may be run continuously for three hours on a running rack to bed-in the bearings to make sure that the shaft rotates freely, before being tested and assembled to a record-changer.

Special precautions are taken in the manufacture of the component parts of the turntable driving system to avoid rejection after assembly. The inside rim of the turntable is driven by a rubber intermediate wheel (see Fig 15) which in turn is driven by the motor pulley. The position of the rubber wheel and the angle of its tension spring must be carefully chosen so that the rubber wheel wedges just sufficiently between the motor pulley and turntable rim to drive without any trace of slip, but it must not wedge enough to cause undue loss of power and, incidentally, to increase rumble.

Should the motor pulley run more than 0.0002 in. out of true on any of its four steps it will cause flutter. This pulley must be interchangeable to accommodate differences in motor speed on various supply frequencies, and the required accuracy of the fit on the motor shaft is achieved by the use of special boring and turning machines developed for the purpose. The bore of the pulley may be "blind," owing to the fact that the step diameter for the speed of $16\frac{2}{3}$ is smaller than that of the motor shaft on which the pulley fits. The speed of $16\frac{2}{3}$ rpm is used for certain special recordings such as talking books.

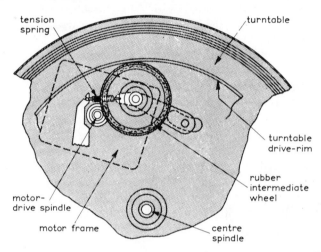

Fig 15. *An intermediate-wheel turntable rim-drive system. The angular relationship with the stepped motor pulley must be correct and the rubber idler must be clean, truly round and of the correct hardness. It is shifted to the correct motor pulley diameter for each speed.*

To give an accurately sized and parallel hole in such a pulley, it is bored to an accuracy of ±0.0001 in., but because of the very close tolerances, a small slot may be broached along the length of the bore to allow the air to escape so that the pulley can be assembled to the motor shaft. All four pulley steps are finish-turned by driving the pulley on a stationary arbor by means of a rubber wheel. When finished, the pulley is locked on the motor shaft by two diametrically opposite screws, which must be equally tightened to maintain the true running of the pulley.

The periphery of the rubber intermediate wheel is ground and then checked on a specially developed electronic instrument for truth and consistency of the resilience of the rubber. This check was found to be necessary in order to reject wheels having soft or hard sections in the rubber, which can cause wow and rumble.

The final link in the driving system is the inside of the turntable rim, which must be round and run true. Most record-changers have pressed-steel turntables and, as it is not always practicable to machine the rim to provide a perfectly true driving surface, great care is taken in the forming of the rim to make sure it is round; the turntable is then held in an accurate centralizing chuck for piercing the centre hole.

On assembly, special attention is given to ensuring that all driving and driven surfaces are parallel to each other. For example, should the rubber intermediate wheel or motor pulley be slightly out of square it would cause what is known as "scrubbing," which would be reproduced as flutter.

The electric motor used to drive the record-changer can be a prolific source of trouble if certain precautions are not taken in its manufacture. It must run smoothly without hum, vibration or mechanical noise, and must have the minimum external magnetic field. It must also be robust and capable of running for long periods without the need for frequent lubrication, and even though it may be enclosed in an unventilated cabinet no over-heating should occur.

The smooth running of the motor is achieved by attention to the finish of the rotor shaft and bore of the bearings, as already described. Hum is avoided by making sure that the stator coils are, in the case of a four-pole motor, electrically in balance and that the rotor is exactly in the centre of the stator bore, this being one reason for spherical turning of the outside of the bearings, as previously mentioned. In addition all rivet- and location-holes in the motor laminations and bearing covers are held to very close tolerances.

Motor vibration, which can be a main cause of rumble, depends upon the true running of the rotor shaft and the balance of the complete rotor assembly. The rotors are assembled on the super-finished shafts, which are then stress-relieved. The outside diameter of the rotor is ground to size on an automatic machine and given a thin even coating of lacquer to prevent rusting. From experience, especially in the export markets, it has been found necessary to give special attention to what appears to be a simple operation, and the procedure adopted for lacquering rotors is worth describing. After grinding, the rotor assemblies are fed on to a ramp which conveys them through an electric oven having a controlled temperature. This drives off all traces of moisture or flux which may be trapped between the riveted laminations. The rotors, while still warm, are rolled over a ramp through a controlled depth of lacquer and passed through another drying oven to bake the lacquer.

The rotor assemblies are then checked for tightness of the shaft in the rotor by applying an end pressure of 250 lb in an air-operated jig, which also checks that the position of the rotor on the shaft is correct. The shaft is then checked for truth by rotating the assembly on its normal bearing locations and clocking the shaft at the point where the pulley will be fitted.

The rotor assembly is now ready for balancing, the accuracy of which is one of the main factors governing the final performance of the motor, especially in the reduction of flutter and rumble.

The rotors, which are dynamically balanced to an accuracy better than 0·031 cm-g on the latest type of automatic balancing machine, are fed into a magazine and are picked up one at a time by an arm which grips the rotor with the aid of a suction cup and places it on "V" bearings, where it is revolved by means of a rotating magnetic field. The amount and location of any out-of-balance is registered by the machine. The magnetic field is then switched off and the rotor is picked up by the arm, placed on a fixture and, if required, recesses are drilled in the copper end-plates at points previously registered by the machine to correct the balance of the rotor. A further position on the machine finally checks the rotor. If it is within the pre-set limits for balance it is passed; if outside the required limits it is rejected.

All the principal shafts on a record-changer run vertically and the thrust bearings on two of them, those for the turntable and rotor, can be a source of rumble. To avoid this the ends of the rotor shaft are lapped to a high finish so that when running on its ball thrust the minimum of friction and noise is generated. If a stationary record spindle passes through the centre of the turntable spindle a ball race must be used to carry the weight of the turntable. A normal commercial type of thrust ball race would be too noisy, consequently specially made races with five steel balls in a plastic cage between two hardened and lapped steel washers are used. The complete ball race assembly rests on a resilient plastic washer to cushion the weight of the rotating turntables from the body of the changer.

In addition to the various techniques developed and used in the production of components for the turntable driving system some of those used to produce the changer mechanism are also of interest.

The first requirement in the automatic handling of records is for the record-changer to drop them individually on to the turntable for playing. This is achieved by means of a stepped record spindle incorporating a lever operated by the record-changer mechanism to push the records off the step one at a time.

There are many types of such spindles, the example described being one of the most popular. The body of the spindle is stainless steel and four milling operations are required to produce the step and slots for the pushing lever and latch. The use of stainless steel was found necessary in order to withstand the abrasive action caused by records dropping and revolving around the spindle. Over a period of time this abrasive effect is sufficient to remove any plated finish.

The spindle assembly must be free from all traces of burr and these are removed from the spindle by brushing, and from the pushing lever and latches by roto-finishing. This latter process is used for removing burrs and sharp edges from practically all levers used on record-changers. The parts are placed in a drum with mineral chippings and water and are vibrated or tumbled for a given period. The method, size of chipping and time depends upon the shape of the levers and the material from which they are manufactured. The dimension limits for the record spindle have to be closely held so that a record on the top limit for thickness will not jam, and two will not drop together should they be on the bottom limit.

The most vital and sensitive mechanism on a record-changer is the automatic trip. This operates when the pick-up reaches the end of a record and engages the mechanism to complete one cycle. The mechanism used is known as the velocity type of automatic trip. As the pick-up approaches the end of playing a record a lever underneath is pushed inward by the pick-up arm. This lever, which has a friction pivot, intermittently engages with a projection on the revolving turntable hub. While the pick-up slowly moves inward, as it does while playing a record, the projection as it revolves pushes the lever away from it. However, when the pick-up reaches the run-out groove at the end of a record it moves the lever inward too far to be pushed away by the projection which then positively engages with it; this action causes the mechanism to engage and complete one changing cycle. The reliable operation of the trip is the basis of the whole mechanical performance of the changer and it must be sensitive enough to operate on a pick-up acceleration over $\frac{1}{16}$ in. within one revolution of the turntable. The action of pushing the lever away must be so light that it cannot be heard via the pick-up, and the

Fig 16. *A high-quality turntable kit for easy home assembly. A flexible precision-made rubber belt is moved by hand to the appropriate motor spindle step and drives the turntable rim.* $33\frac{1}{3}$ *and* 45 *rpm speeds are provided as being suitable for the vast majority of modern records. High-precision bearings and parts allow very low wow and rumble figures to be achieved.*

maximum side pressure which can be tolerated on the pick-up arm to operate the trip is about 0·25 g. It also has to cover a wide operating radius to accommodate all the types and sizes of records now available.

To achieve all this, the associated levers must be free from burr so that they move freely, and special lubricants are used, as ordinary oil would congeal and prevent operation. Clean and careful assembly is absolutely essential.

The final testing of record-changers calls for special instrumentation and most of the equipment used has been developed specially. Every changer is checked for correctness of all electrical and operational functions at inspection stations at the end of the assembly lines. Among the tests given are those for wow, flutter and rumble, all of which require the use of special test records and sensitive instruments. In addition there may be twenty-five operational functions to be checked on each record-changer and the failure of any one will cause it to be rejected.

Every changer should be checked for correctness of the following functions, but not necessarily in the order given—

1   Dropping one at a time, a stack of mixed 7 in., 10 in., and 12 in. records.
2   Operation of record-size selector mechanism.
3   Switching off after last record has played.
4   Height of pick-up lift, adjusted if necessary.
5   Lowering position of the pick-up using a special test record giving correct landing radii.
6   Operation of automatic trip when pick-up reaches end of record.
7   Stylus pressure; adjusted if necessary.
8   Operation of pick-up muting switch, if fitted.
9   Effectiveness of switch-click suppressor.
10 Lateral and vertical freedom of pick-up arm. Point friction must be of a very low order for modern low-weight cartridges.
11 Correct operation of speed-change control; check that the intermediate wheel is in the centre of the appropriate motor-pulley step.
12 Use special stroboscopic discs to check that, while the pick-up is playing a record, all speeds are within ± 2 per cent of nominal.
13 Operation of "off," "manual" and "auto" control.
14 Operation of pick-up-arm bias compensator.
15 Angular location of record spindle.
16 Satisfactory operation when using large record spindle for dropping 7-in. records having 1·5-in. diameter centre hole.
17 Absence of overall mechanical noise.
18 Satisfactory operation on 10 per cent below minimum rated voltage.
19 Motor current consumption.
20 Phasing of stereo pick-up connections.
21 Angular position and freedom of record overarm.
22 Cartridge output and channel separation on stereo when a cartridge is supplied.

23 Insulation between motor and pick-up circuits and frame of changer.
24 Check setting of links in voltage-changeover terminal block and marking on motor (corresponds to its voltage and rating).
25 Finish and completeness.

Fig 67 shows the Garrard AP 75 semi-automatic high-quality single-record-player integrated with an amplifier system and plinth.

It should be noted that there are available basic transcription or other high-quality turntables without any refinements such as automatic lowering or stop facilities. This saves money and the value may be concentrated on the essential motor and turntable bearings and finish. The additional cost of adding a high-quality tone-arm must, however, be added on, and the need for careful manual lowering etc of the pick-up has to be considered.

The system as a whole
The gramophone disc-recording system is capable of giving extremely faithful reproduction provided that the system is designed and considered as a whole. It has been demonstrated that it is possible to record and reproduce television pictures on gramophone discs and to reproduce them with a pick-up basically similar to those in normal use at the present time. The pictures were necessarily not transmitted at the full rate possible with the normal electronic television systems or with rotary head cross-scanned television magnetic-tape-recorders, as these demand a frequency response extending to 3 Megahertz or more. Nevertheless, a rapid sequence of high-quality pictures was produced on gramophone records, which brings home the remarkable potentialities of the system. Improved disc TV recording systems are now being demonstrated.

As the gramophone disc and pick-up combination is predominantly mechanical in operation, it is likely to suffer a drastic loss in quality unless it is correctly understood and meticulously maintained by users, particularly as regards cleanliness. Top-class recording companies can be relied upon nowadays to cut their master records on very high-precision machines, and to comply with the international standards, with due regard to such factors as correct groove profile and surface finish, avoidance of excessive modulation, particularly avoiding excessive values of peak acceleration of the cutter tip.

It is obviously of paramount importance that the pick-up should at all times trace accurately the groove modulation. The greatest single source of deterioration of quality in gramophone record reproduction lies in the failure, for one reason or another, to achieve accurate tracking or tracing of the groove modulation by the pick-up.

These errors in following the groove modulation fall basically into two categories—

(*a*) Those errors due to the fundamental geometry of the groove, stylus and pick-up generally. The study of these limitations is usually classed under the heading of "tracing distortion." Various measures can be adopted to minimize this trouble, both in the recording and the reproducing processes.[7]

(*b*) Errors due to faulty record maintenance and inadequate cleaning; also to faults in pick-up design, maintenance and wear of stylus and record etc.

The whole subject is extremely involved and it is really necessary to have a firm grasp of the fundamentals in order to operate and maintain your gramophone equipment, records and pick-up properly, so as to realize the full capabilities of modern recordings.

The first essential is to appreciate the various properties of the groove and record materials. The recording process is an extremely involved branch of precision engineering which will merely be outlined here in so far as it has a direct bearing on the properties of the groove and thus affects reproduction.

Properties of monophonic and stereophonic record grooves

Record masters are cut by a chisel-shaped "Vee" profile sapphire cutter which is very carefully ground to the desired groove section and is highly polished. Modern cutter heads are multi-winding moving-coil instruments designed to drive the cutter over a wide frequency range, both vertically and laterally or in any desired combination of these motions, so as to produce stereo or mono recording. Feed-back windings provide inverse voltages which are injected into suitable points in the driving amplifiers, so as to flatten the frequency response and to reduce distortion in the amplifier-cutter head combination. Hard-metal record-stampers are produced by successive plating operations, starting from the master. Many programmes are originally recorded on tape and, in this case, the maximum record-playing time can be achieved by driving the recording traverse lead screw from an electrical system fed from an advance tape replay-head, which enables the record groove spacing to be pre-set to suit the signal amplitude about to be recorded. Sophisticated electrical devices such as non-distorting volume limiters and devices designed to avoid unduly high recorded values of acceleration are now widely used in recording studios. In addition, it is possible, by using the "Dyna-groove" type of technique devised by RCA,

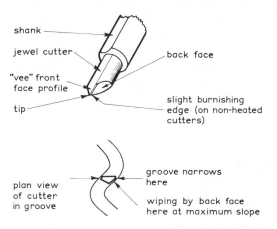

Fig 17. *The action of a gramophone recording cutter when producing the groove. The chisel-ended cutter moves both laterally and vertically to produce a stereo recording. The groove angle narrows at the centre point, where the velocity is a maximum and the absolute value of slope is limited by the shaving of the groove wall by the back angle of the cutter rear faces.*

to reduce the basic amount of tracing distortion, which is due to the finite groove and stylus sizes which have to be used. This technique is only fully effective if the results of record wall elasticity and the exact pick-up stylus size are accurately known.

It is necessary to cut a very highly polished groove, without any tendency to produce rough edges by "tearing" of the swarf (the removed ribbon of material) along the top edges of the groove. In order to engrave the higher audio-frequencies without loss, the cutter must be very sharp with a correct back angle behind the cutting surface. A dead-sharp cutter is liable to leave a rough groove, unless the stylus is heated to the correct temperature by a small DC heating coil wound on the top part of the cutting sapphire. In many cases the cut swarf is blown clear of the groove by an air jet and is then removed by suction.

The plastic material selected for record-pressing must have a good thermoplastic flow, a high surface finish, low friction and good mechanical and thermal stability. Vinyl blended with certain additives meets the requirements fairly well, and is almost universally used today. Its main limitations are that it is somewhat elastic, it is subject to some variation of properties with temperature changes, it is electrostatic (thus picks up dust) and it may be subject to warping or "cold flow." Some of the additives used today are designed to minimize these shortcomings but, for the most part, it is incumbent on the user and the pick-up designer to play and store records in such a way that the best results can be obtained from the present material.

New plastic materials are continually being investigated and when an improved record material becomes available at the right price it will doubtless replace the present materials used.

The record groove

Both mono and stereo recordings are made with the same standard microgroove whose section is typically defined—

Width of groove = 0·0025 in.
Width of inter-groove land = 0·0015 in.
Depth of groove = 0·0013 in.
Radius of groove bottom = 0·00015 in.
Angle between groove walls = 90°.
Groove width + land width = 0·004 in.
Number of grooves/inch = 250.
Maximum recorded amplitude = 0·00075 in. (up to 0·002 in. may be recorded when automatic groove widening is used during recording).
Recorded wavelength at 10,000 cycles/second—
(i) 0·003 in. at outside of disc.
(ii) 0·0015 in. at inside of disc.
Monophonic modulation is lateral.
Stereo modulation is +45° and −45° on each side of the vertical for the two channels.

The dimensional accuracy of the groove profile and the surface polish of the walls must be of a very high order indeed. This is brought home to us if we realize that the smaller audible amplitudes reproduced by the pick-up may be only a few millionths of an inch in magnitude (ie a few micro-inches). A surface with irregularities less than a micro-inch normally represents a very high polish.

The leading recording companies take steps to check the dimensions and surface finish of their recording cutters at frequent intervals with high magnification shadow-graphs and microscopes, so as to ensure that the grooves in their master records are smooth and accurate.

It was discovered very early in the history of recording that the groove had to be cut with an almost dead-sharp chisel-ended cutter, which removes the surface of the master record in a smooth continuous thread of swarf. Attempts to emboss a groove into the surface of the record by a spherical-ended stylus have not been successful for high-quality recording, partly because of the large amount of force required to be generated by the cutter head and also because the walls of material thrown up on each side of the groove have proved to be a source of trouble in the record-manufacturing processes.

The reproducing pick-up stylus must be a highly polished sphere *or* ovoid of exactly the correct radius, in order that it may fit accurately in the low part of the record groove without either actually touching the groove bottom or riding on either or both top edges. Styli of the normal outside permitted manufacturing tolerance limits will fit in an international standard groove. A careful study of the geometry of the motion of the V-shaped cutter, the moving groove and the spherical pick-up stylus shows some interesting and important factors—

1 The cutter crosses the centre axis of the groove at its maximum modulating velocity twice per cycle. The groove can be regarded as moving at an angle to the plane containing the cutting face and thus the angle of the groove is narrowed twice per cycle. This means that the reproducing stylus is squeezed upwards by a small amount. This affects the design of pick-ups in that the vertical mechanical impedance must be kept low so as to avoid excessive forces on the groove wall. Also, in a mono pick-up, if the transducer mechanism produces an output due to this vertical "pinch effect," spurious signals appearing as noise or distortion will be generated. In stereo recording the pinch effect is an inevitable slight complication which can cause

distortion and inter-channel cross-talk at high values of modulation velocity.

2 It is necessary for both the cutter and the pick-up stylus to have a minimum tip radius of a few tenths of a thousandth of an inch in order to ease manufacture and increase durability. However, if the groove has a sine-wave excursion, the centre of a

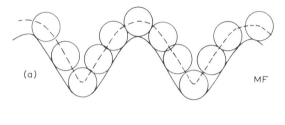

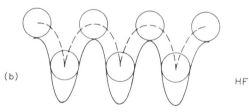

Fig 18. *The tracing of the groove by a hemispherical-ended stylus. A sinusoidal stylus excursion will not be obtained from a high-frequency groove produced by a sinusoidal cutter excursion, giving rise to "tracing" distortion. This is minimized by the use of an elliptical-ended pick-up stylus in which a very small side radius is used. An additional technique is to "pre-distort" the groove, as cut, in the opposite sense, so that a stylus with a given spherical radius will generate an undistorted wave. This assumes that the stylus reaction force produces no appreciable elastic deformation of the groove wall.*

spherical stylus tracking the groove will not trace out a sine wave; this is illustrated in a very simplified way in Fig 18. There is some cancellation of the tracing distortion for lateral mono reproduction owing to the fact that one wall of the groove to some extent compensates for the errors generated by the other wall. This is not so in stereo recording and the "reproduction-correlator" or "Dynagroove" recording systems have therefore been evolved. As previously mentioned, these consist in adding a carefully predetermined amount of inverse distortion to the recorded wave as a result of making certain assumptions as regards pick-up stylus tip radius, record elasticity and plasticity, pick-up mechanical impedance and stylus tip mass etc. This type of system is by no means easy to apply owing to the unknown pick-up factors involved in different reproducing systems. Record companies may rely on keeping tracing distortion within

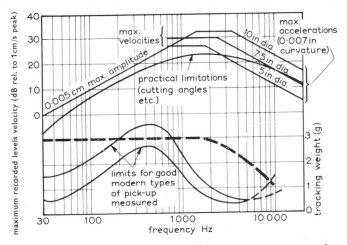

Fig 19. *The maximum permitted levels recorded on gramophone records in terms of the "tracking ability" of practical high-quality pick-ups. (The dashed line shows the maximum permitted weight.) A lack of sufficient lateral compliance causes the stylus to climb the inner wall at peak excursions, whilst an excess of mass acting at the stylus tip may cause the stylus to climb the outer wall at peak excursions, where the acceleration is a maximum. The stylus may also fail to maintain contact after the peaks of vertical acceleration on stereo recordings. The broken line shows the maximum tracking weight ideally allowable if deformation of groove walls is to be avoided at the frequencies shown for the maximum recording levels encountered.*

reasonable bounds by avoiding excessive modulation levels, particularly at high frequencies. This can be done automatically to a certain extent by using "limiter" amplifiers with appropriate characteristics. Fig 19 shows maximum recorded modulation levels for standard recordings and styli as well as some of the relevant pick-up properties.

Stylus and arm angular effects

Several of these geometrical effects are involved in recording and reproduction. One is the so-called "tracking angle," which is the angle between the vertical plane in which the stylus tip swings and the radius from the disc centre. Recording machines have an accurate lathe-type radial traverse slide, but most domestic pick-up arms swing about a vertical pivotal axis. This inevitably involves some tracking angular error, which causes a certain amount of distortion of the reproduced waveform. This can be minimized by mounting the pick-up cartridge at a suitable offset angle to the axis of the arm (usually between 15° and 30° according to arm length etc). The vertical operating angle of a stereo-recording cutter is usually set at about 15° positive angle, as opposed to the negative angle of the old-style acoustic gramophones. If the pick-up stylus is not set at a similar angle, some distortion will occur. There is now a move to standardize the vertical recording angle. It is generally accepted that distortion is imperceptible if lateral tracking error is held to less than 5° and the vertical stylus angle is within ±10° on +15°.

The above considerations illustrate the way in which the mechanical limitations and geometrical errors of our recording and reproducing devices introduce noise and distortion into gramophone record reproduction. It would obviously be better if infinitely sharp tips could be used on the recording cutters and pick-up styli. In practice the minimum realizable values are about 0·0001 in. tip radius for the cutter and about 0·0002 in. to 0·0003 in. for a spherical-ended or ellipsoidal pick-up stylus. These values, together with the high polish and good surface finish which can be imparted to record grooves on modern vinyl pressings are in fact capable of giving extremely good-quality sound reproduction. It is other factors such as the mechanical properties of the pick-up moving system and the record material, record warping, dirt etc that are likely to set the practical limits at present to gramophone reproduction.

The record material and the stylus

The plastic material used for record-pressing is, of course, not infinitely rigid and it will be deformed slightly at the points of contact of the stylus tip. Any yielding of the groove walls means that the mean path traced by the stylus centre will be slightly distorted. The distortion depends on the force between the stylus and the groove. This is primarily dependent on the mechanical impedance of the pick-up system referred to the stylus tip and the downward weight which has to be applied to the pick-up in order to keep the stylus continuously in contact with both groove walls at all points of any modulation cycle which is recorded. The pinch effect and the vertical components of 45°/45° stereo recording mean that the vertical mechanical impedance characteristics of the pick-up system must be considered as well as the horizontal components. The effects of the indentation of a flat surface by a static load applied to a spherical indenter have been fairly extensively studied. The dynamic effects present when a pick-up stylus rides the two walls of a modulated groove with continually varying groove-wall curvature and instantaneous forces are very much more complicated and are very difficult to analyse. Up to a certain load, the record-groove wall surface will be elastically deformed, that is, it will spring back to its original form without any damage after the stylus has passed. Above a certain critical value, given by the yield point of the record material, plastic deformation represented by permanent surface or sub-surface damage and indentation will occur. In other words, the pick-up will leave marks on the groove walls and a greater or lesser degree of record wear will take place.

An enormous contribution to our knowledge of plastic and elastic record-wall deformation is due to two British experts each working independently along his own particular line. The first was Cecil Watts, who first perfected the art of record-groove photomicrography and analysed an enormous number of records of all kinds, both played and unplayed. The second was John Walton,[7] who, in addition to design work and theoretical studies of pick-ups, arms and grooves, developed a remarkable technique of taking electron-microscopic photographs of parts of record-groove walls at magnifications of up to 20,000 times. These show up in a most revealing manner the effects of instantaneous plastic deformation of the groove walls at points in the modulation cycle and make it clear that the downward force and the mechanical impedance of the stylus load must be kept to a very low value if plastic deformation and wear of the record walls are to be avoided. It is, of course, desirable to avoid even any appreciable elastic deformation of the groove wall. The pick-up mechanical impedance is largely governed by the restoring stiffness together with the effects of pick-up arm resonances at low frequencies, by the system resistive damping at middle frequencies and by the effective mass at high frequencies, all values being referred to the stylus tip. It is particularly difficult to keep the moving mass to a low enough value, as the mass of any but a small stylus jewel rondel is liable to be excessive, without allowing for the other necessary moving parts of the transducer mechanism such as a moving magnet or armature, coil, crystal etc. The effect of any appreciably excessive mass may be to "wipe off" the high-modulation from the record and cause a loss of brilliance on subsequent reproduction. Reference[7] gives a wealth of detail and references on this subject, together with some most revealing electron-microscopic photographs, showing for example, that in one case the effect of 300 playings with a 0·6-milligram stylus-tip mass was merely a polishing of the slight score lines and rough-nesses left on the groove walls by even the highly polished cutter used for recording, whilst the effect of a 3-milligram tip mass was to obliterate the original high-frequency recording and to leave a distorted indentation of its own on the groove wall. The achieve-ment of very low values of stylus-tip mass taxes the skill of the pick-up designer and requires the use of artifices such as the mechanical leverage action between the tip jewel on the stylus arm and the transducer element or armature.

Pick-up design and performance specifications
The mechanical impedance of any pick-up moving system de-pends on the size, weight, layout and mounting of the various parts concerned. The analysis and development of these parts are a complex and exacting task. This fact probably accounts for the fact that there are as yet relatively few really first-class designs of pick-up produced in the world.

Pick-ups are nowadays mainly electromagnetic or piezo-electric (crystal) types. The first category usually depends on the induction of a voltage in a coil by either a moving armature or a moving magnet. Both types of element may be made very small, whereas a moving coil tends to be larger and more difficult to mount. A piezoelectric crystal or ceramic element tends to be relatively large but the high transduction efficiency allows the use of a flexible coupling between it and the stylus, thus affording mechanical decoupling of the transducer mass without encounter-ing an excessively small output voltage level from the pick-up. The effects of temperature and time on the various plastic mount-ings, stylus arm and damping elements etc must be considered by the maker so as to ensure a consistent performance and a satisfactory life for the pick-up cartridge under all conditions likely to be encountered.[7,11]

We have mentioned the three régimes on the frequency scale in which the stiffness, resistance and mass of the moving system predominate, at low, middle and high frequencies respectively. The effect of any mechanical resonances may be to increase or reduce the impedance according to whether they are effectively "parallel" or "series" resonances when referred to the stylus tip, in an equivalent electrical circuit analogy. The placing and damping of these resonances is thus as important as the realization of suitably low values of stiffness and mass at the stylus tip. The applied downward weight offsets the reaction due to the mechanical impedance. The restoring stiffness reaction at low frequencies tends to cause the stylus to climb the near wall (ie that closest to the centre line) of the modulated groove at its peak excursion from the mean position, whilst the mass reaction at high frequencies causes the stylus to attempt to climb the far wall at peaks, owing to the kinetic energy acquired at the maximum velocity (centre crossing) points. This causes a tendency for the outward motion to be continued, with the result that the outer wall may be mounted. In both cases, mis-tracking will occur.

The smallness of the downward weight required for the maintenance of correct two-point stylus-groove contact at *all* points of the cycle is a direct measure of the extent to which the mechanical impedance of the system has been kept down to a suitably low value at any given frequency. The "break-point" at which intimate contact is lost is manifested as a sudden increase in distortion. The procedure for testing for the break-point is described later.

Stereo recording and reproduction

We have seen that the standard form of stereo gramophone recording is a two-channel 45°/45° system in which each groove wall in effect moves independently at 45° to the vertical groove centre line to give the left- and right-hand channel signals. The recording is made with a cutter head with two windings which drive a normal cutter with a composite motion which corresponds to the instantaneous resultant of the two channels. When the signals from the two channels are "in-phase," this resultant motion is represented by a lateral motion of the groove. When the two signals are in "anti-phase," the groove undulates vertically.

In all normal recorded works the in-phase signals are stronger than the anti-phase signals and the lateral excursions predominate.

The stereo pick-up must thus have two independent sensing systems and a mechanical system capable of tracking the groove correctly in both horizontal and vertical directions. Thus it requires a lower vertical mechanical impedance than is the case for a mono pick-up. This poses difficult design problems on the

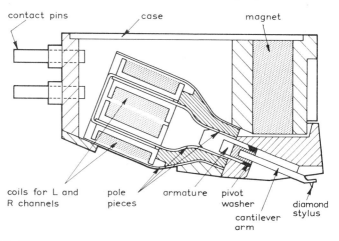

contact pins   case   magnet

coils for L and   pole   armature   pivot
R channels   pieces   washer   diamond
                                      cantilever   stylus
                                        arm

Fig 20. *A cross-section of a modern high-quality "induced-flux" type of stereo magnetic pick-up. Flux from the fixed permanent magnet is fed to the two pairs of poles, which are mutually set accurately at right angles and carry the right- and left-hand stereo windings, by a miniature ferrite armature driven by the cantilever and stylus, the whole moving system being pivoted at its centre of gravity in a lightly damped elastic mounting ring.*

transducer unit mechanical layout and suspension but, fortunately, the low-frequency signals in particular tend to be in phase in the two channels and thus do not require such a high degree of vertical compliance as for lateral. Typical values might be—

Lateral pick-up compliance at stylus = 20×10^{-6} cm/dyne. Tip mass less than 1 milligram.

Fig 20 shows the construction of a representative modern stereo pick-up of high quality in which both lateral and vertical mechanical impedance is kept at a low value referred to the stylus tip. Different models offer a range of cartridge constants above and below the values above. The correct unit should be selected for a given arm and turntable, bearing in mind that very high-compliance low-weight cartridges require near-perfect conditions in order to realize their full performance.

Checking procedure for correct pick-up tracking

Assuming that all avoidable side-thrust due to sources such as excessive arm-pivot friction, inaccurate turntable levelling and record warp has been eliminated, and also that offset angle bias torque has been compensated in the pick-up arm design (page 73), then the mechanical impedance characteristics and groove tracking may be checked by the following method. The downward weight on the pick-up head is gradually reduced until distortion just becomes perceptible when heavily recorded passages are being played. The onset of severe distortion is quite sudden as the stylus tip loses the proper two-point contact with the groove and commences to climb the groove wall. Listening checks of this may be supplemented by examination of the reproduced waveform on an oscilloscope. The "break" represented by the onset of bad groove tracking should be clearly visible. Frequency records may also be used.

If there is excessive high-frequency distortion, excessive tip mass at the stylus point or a high-frequency resonance may be indicated.

Procedures such as the above can, with experience, give quite good assessments of the quality of a pick-up and whether it is likely to meet the criteria already outlined on groove deformation and wear.

The maximum recorded peak values of displacement amplitude (x), groove velocity ($\dot{x}$) and acceleration ($\ddot{x}$) encountered on standard lateral microgroove recordings are indicated in Fig 19 (page 65).

If the values of the pick-up lateral mechanical constants are referred to the stylus tip, the value of the stylus reaction force at the groove wall can be calculated to a first order of accuracy. It must be assumed that the stiffness of the stylus-restoring (-centring) system is the dominant factor at low frequencies, but that the stylus-tip mass is dominant at high frequencies. The stiffness-reaction force is given by the lateral stiffness constant times the peak amplitude. Thus, a compliance constant of 5×10^{-6} cm/dyne at a peak excursion of 0·005 cm (0·002 in.) produces a lateral reaction force of 1000 dynes or approximately 1 g. The vertical force required to maintain groove contact will be equal to the lateral force for groove walls inclined at 45° to the vertical, but extra weight is advisable as a factor of safety to allow for record warp etc.

At high frequencies, the peak groove acceleration may be as high as 1000 g. Thus, a stylus-tip mass of 2 milligrams will cause a reaction force of 2 g.* These simplified calculations may be in error owing to the effects of resonances, as well as the effects of damping resistance, which may increase the reaction forces considerably. The resonance between the pick-up arm mass and transducer restoring stiffness occurs at a very low frequency, and if it is not damped it may cause bad tracking and may exaggerate the effects of rumble. Warped records, poor levelling of the turntable, side-thrust due to pick-up-head offset angle and other effects can increase stylus reaction forces and cause bad tracking.

Thus, in order to allow a reasonable factor of safety, a downward tracking weight of up to 3 g may be needed under adverse conditions, even when the pick-up has almost ideally low tip mass and high suspension compliance. It is also necessary for the pick-up movement to have a low vertical impedance so as to allow for the twice-per-cycle pinch effect lift due to groove narrowing. Unfortunately, there does not appear to be much latitude in regard to pick-up constants and downward weight. The tip radius should be as small as possible so as to minimize tracing distortion, but it would appear that we are then limited to something like the mechanical impedance and downward weight given above in order to avoid the possibility of plastic (permanent) damage to the groove wall at points of high acceleration. In practice, taking all factors into account, including the effects of tone-arms, record-changer mechanisms etc, the manufacturer has to arrive at some compromise between tip radius and the other factors involved. It is thus necessary to choose a cartridge with constants appropriate to the duty involved.

Checking for stylus wear and damage

We have stressed the need for a precise spherical or ellipsoidal highly polished sapphire, ruby or diamond tip for the pick-up stylus. With suitably low tracking weights, many thousands of record playings should be possible before any significant wear of

---

* See note on page 223.

a jewel tip occurs. A good diamond point should be regarded as substantially permanent for a low-weight pick-up. It is possible, however, for some stylus tips to fracture or to become badly worn in the course of time. Severe tip damage may be audible as noise or distortion and the groove may be scored and left dull in appearance.

Fully satisfactory visual checks unfortunately require a microscope or shadowgraph of higher magnification than is normally available in the home or in any but the larger shops (about 200 × magnification is required for 0·5-mil tip radius).

It is thus best either to use a diamond tip or to keep count of the number of playings by sapphire tips if these are used and to replace by a new tip after a great many records have been played. If the tip is appreciably worn, a change to a new tip will give an immediate improvement. Small microscopes of about 50 × magnification are used to give some guide to stylus condition in home use.

Needless to say, the correct replacement tip and attachment (usually a mounted cantilever spring) must be obtained for the model of pick-up and it must be carefully fitted after removing any dirt or fluff which may have accumulated inside the pick-up mechanism.

STYLUS DOWNWARD-WEIGHT CHECK

It is important to check the downward weight on the pick-up stylus point and also to check the vertical and lateral arm pivots for freedom of movement. A spring stylus pressure gauge reading from 0·1 to 5 g is suitable. Many modern arms have a calibrated micro-adjustment of the tracking weight and the bias adjustment.

Pick-up arm design

The arm is a very important element of the system and must fulfil several functions—

1 It must provide a light, rigid and resonance-free support and attachment for most standard types of cartridge.
2 It must be freely pivoted in both horizontal and vertical planes, without any shake or lost motion in the pivots. Frictional force at the stylus point due to the pivots should not exceed 0·5 g. The best modern arms give much lower figures than this.
3 The correct downward weight to give accurate tracking for

any particular cartridge must be accurately adjusted. An adjustable counterweight is preferable to a spring.

4 The arm must angle the cartridge so as to minimize the tracking angle error for the particular mounting distance specified between the record centre and the arm vertical pivot centre. A typical offset angle is 15°.

5 The use of an offset angle, whilst minimizing tracking angular error, inevitably introduces a torque tending to pull the pick-up towards the record centre. It is desirable to compensate for this, and the best arms incorporate a drawstring and small weight, a magnetic bias, specially angled tension springs or other devices which can be set exactly to balance the offset torque, ie to give side-thrust or bias correction.

Certain points in the design of arms are of interest. The arm itself should be rigid and well-damped without being heavy. If it is flimsy or of too thin a section, torsional resonances may be present and, in extreme cases, these may become sufficiently strongly excited to upset the response of the pick-up or to add a ringing "coloration" to any background noise. Substantial aluminium rods, mouldings, die-castings or hardwood are successfully used for arms. Pivots are a particular problem. They should be fairly close to the record surface to avoid back-and-forth movement of the stylus which might cause wow on slightly warped discs. The vertical and horizontal pivots should preferably be as close together as possible or the effect of any slight errors in levelling of the motor board and turntable will throw the arm out of balance and exaggerate the side-thrust produced. The ideal type of pivot is a polished hard single point or knife edge which will thrust any incidental dirt particles aside and whose effective frictional radius relative to its centre is very small, thus minimizing the effects of pivot friction.[12]

We have mentioned that the effective mass of the arm resonates with the pick-up transducer suspension compliance at a low frequency, generally between 10 Hz and 40 Hz, in both the horizontal and vertical directions. This resonance determines the lowest frequency reproduced and, unless it is properly damped by some mechanical resistance around the transducer suspension system, rumble may be exaggerated and an excessive mechanical impedance presented at the stylus point, possibly causing bad tracking or shake. If the pick-up and arm system is correctly designed as a whole, the low-frequency response will be flat for both lateral mono and stereo recordings down to below 20 Hz (ie below audibility) with a sharp cut-off below this point. This is effectively a "built-in" rumble filter with good impedance and response characteristics.

Figs 14 (page 50), 66 (page 207) and 67 (page 208) show good modern pick-up arm designs mounted on motor boards.

Record care and cleaning

NEED FOR CLEANLINESS AND FLATNESS

We have seen from the previous sections on the record groove that quite small traces of dust and contaminant will be sizeable compared to the groove and stylus tip, and thus can cause loud "pops" in the case of discrete particles struck by the stylus and may cause distortion and bad tracking in the case of "sticky" surface contaminants. Even small accumulations of dirt will cause bad tracking and distortion in lightweight cartridges.

It is obviously essential to keep records as clean as possible and to replace them in their covers as soon as possible after playing.

Warping is another grave defect which can arise only too easily if records are not stored and used correctly. Anything more than a very small amount of warp will cause serious wow on reproduction because of the cyclical "stretching" of the track and the varying stylus-groove geometry introduced during rotation.

Records should not really be left in flat horizontal piles, as is often done, because any irregularity in the stack or on the surface below a record tends to cause the record to go out of flat because of plastic "cold flow." Leaving records in the sun or on top of a warm amplifier case is, of course, a certain way of producing bad warping. Once warped, it is very difficult to flatten records. It would strictly be necessary to place the record between two flat plattens or sheets of plate glass and to heat the assemblage up to just the correct temperature, cooling everything down again before removing the record from the plattens. However, storage on edge under light pressure helps correction.

RECORD STORAGE

It is best to store records in a cool clean cabinet, on edge, with a slight lateral clamping action to hold the stack of records in their cases gently but firmly erect without any tendency to lean. The "toast-rack" type of bent wire or slotted record rack should really only be used as a short-term stand while waiting to play records.

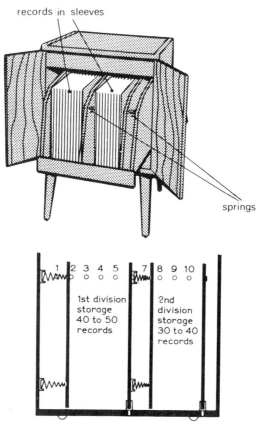

Fig 21. *A new type of edge-storage system which houses the records vertically under a light spring pressure which prevents and helps to remove warp, one of the major hazards to which records are subject when not in use. Moving clips and partitions decrease or increase storage space by* 1–10 *records. Below, a plan view.*

Figs 21 and 22 illustrate commercial systems of record storage on edge in a case or a cabinet where the records may be kept clean. The desired sideways pressure is maintained in the first system.

Record "grips" or handling pads are also sold to prevent finger-marking records (by Highgate Acoustics Ltd, London, W1).

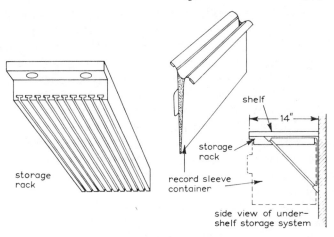

Fig 22. *A convenient and safe system of record storage in automatically sealable transparent plastic pockets which are suspended from a multi-track unit into which they slide.*

TYPES OF RECORD CONTAMINANT

We have noted the need for extreme cleanliness of record grooves, but it is as well to know as much as possible about the subject before attempting any groove-cleaning process, as the wrong treatment may make matters worse.

Records should not be fingered or blown on. Nor, in general, should they be polished, rubbed or washed except with the correct equipment and media.

Record contaminants have been found commonly to consist of (*a*) particles of dust, grit and fluff, (*b*) deposits left by soot from domestic fires, household sprays (when used), solid residues from impure antistatic and washing solutions where these have not been based on properly distilled and de-mineralized water, etc, (*c*) oily deposits from cooking vapours, tobacco smoke, diesel smog etc.

When one realizes the minute size of the irregularities which the stylus can sense on the groove wall (less than the wavelength of light) it is obvious that vestigial traces of oily deposits from the atmosphere may trap minute dirt particles firmly, so that these are forced into the groove wall by the stylus instead of being pushed aside. Even if the foreign matter is subsequently washed away, an audible click may be reproduced from the small crater left in the groove wall.

It is obvious that the less records are exposed to the atmosphere, whether being played or not, the less the risk of all types of contamination. Positive measures aimed at maintaining record cleanliness consist of "dust-bug" types of cleaner which track in the groove whilst it is being rotated for playing, and specialized record-cleaning machines or treatment kits which scientifically clean dirty or contaminated records. Velvet pads with the correct bristles are also used.

The extra pleasure and increased life obtained from really clean records emphasize the value of taking proper steps to achieve and maintain cleanliness.

THE DUST-BUG RECORD CLEANER

This device is produced by Cecil Watts Ltd, the original suppliers. Basically, it consists of a forwardly inclined brush of about two dozen nylon bristles with carefully-pointed ends of the correct tip radius to scour the bottom and sides of the grooves. They are mounted on a light freely pivoted tracking-arm ahead of the pick-up. A total tracking weight under 1 g is possible and a trace of antistatic fluid with a distilled de-mineralized water solvent may be applied via the bristles. It is really desirable that the particles of contaminant collected from the record groove should be removed at once, before they again lodge in the groove.

RECORD-CLEANING MACHINES AND MATERIALS

These range from more elaborate devices based on the "dust-bug" to ambitious routines which apply groove scrubbers, suction mops, spinning polishing brushes and controlled-heat driers. The record should be left in a completely clean condition with the merest trace of antistatic surface film to inhibit the acquisition of further electrostatic charges in the future.

Unless records are given a measure of antistatic treatment it is impossible to prevent the attraction of dust, and all users should apply a scientifically designed and formulated antistat and record-cleaning routine. Once treated and carefully stored, a light wipe with a special clean pad will usually suffice at each playing.

A stylus-cleaning brush should be used with care at each playing to clear the cartridge stylus tip of any dirt.

The "Zerostat" piezoelectric pistol is now made. It releases a stream of + or − charged ions which render records largely antistatic when the trigger is pressed.

4

Magnetic tape-recorders

The recording medium is a long strip or reel of plastic tape coated with a magnetic material. The tape is carried past the erase-, record- and replay-heads by the tape-transport equipment on the tape-deck or machine. It is taken from the supply reel and spooled on to the take-up reel during either the recording or the replay process, rewinding being necessary in each case before another recording or replay operation.

The heads consist of electrical windings on a ring core of Mu-metal or ferrite with a small accurately formed gap across which the magnetic-coated side of the tape passes. The precise design of each of the three types of head has been established as a result of long experience and the manufacture and testing of the tape is a highly sophisticated process. An understanding of tape-recording involves a study of the magnetic properties of the tape coating and the interaction between the magnetic system of the heads and the permanent magnetism established on the tape.[8]

Magnetic recording tapes

Magnetic recording was originally carried out on metal magnetic strip or wire, but about 1940 Pfleumer in Germany made paper- and plastic-based tapes coated with a dispersion of particles of ferromagnetic material. Modern recording tapes have been directly developed from these experiments. Until about twenty years ago, the magnetic coating consisted of small spherical particles of ferric oxide (jewellers' rouge), Fe_2O_3. This material is normally non-magnetic, but was specially treated to produce magnetic properties. These earlier tapes had to be run at 30 in./sec in order to get a good frequency response. Ferric oxide particles were later produced in a long, needle-shaped form, rather than in the original spherical form. These produced tapes with much better magnetic properties (higher coercivity and higher remanence). Recently, chromium dioxide and other higher coercivity tapes have been introduced to improve quality on narrow cassette or cartridge tapes which run at low speeds. Increased bias and erase currents and different equalization are needed for these tapes.

The plastic materials mainly used for the tape base are acetate, PVC and polyester. Acetate was the original material used as it is strong, does not stretch much, and the coating adheres well. It may, however, become brittle with age and it may wrinkle up if it absorbs too much moisture. PVC and polyester are less affected by ageing and moisture, but may stretch more readily, thus causing changes in pitch of a recording. The extra strength of a polyester base is useful for the thin extra-long-play tapes now available.

Standard tape widths are now slightly under $\frac{1}{4}$ in. (0·244–0·248 in.). The following table shows the tape classifications now produced commercially. Available playing time is from a few minutes to over 12 hours, depending on tape speed, spool size and the number of tracks successively traversed.

CATEGORY	THICKNESS (OVERALL)
Standard Play	0·002 to 0·0018 in.
Long Play	0·0016 to 0·0013 in.
Double Play	0·0011 in.
Triple Play	0·0007 in.

A number of other properties of magnetic tape are of great importance and need to be carefully controlled by tape manufacturers and users. Some of these are—

1 "Cupping," or the tendency for tape to develop a concave curvature across its width with the coating inside.
2 "Bias" or "Curl." These are waviness of the tape edges or deviation from straightness. Unless held within limits, these conditions may cause variations in output or speed "flutter."
3 Poor edges. A tape which has been badly slit to width during manufacture may have frayed edges which can shed pieces of coating or lead to tearing.
4 Coating quality and adhesion. The density of the magnetic particles should be as high as possible and as even as possible, the surface polish should be very high, so as to reduce head

wear and friction, and the coating thickness should be even and its adhesion to the base should be reliable. If these points are carefully watched by the manufacturer, the tape should have a good response and signal/noise ratio, free from "drop-outs" or momentary gaps in the signal output.

5 "Blocking" or "Sticking." Turns of tape tend to hold together on the spools because of electrostatic charges or gumminess of the surface, causing snatching and wow during playing. Friction or stickiness can also cause squealing or vibration of the tape at the guides. Care should be taken to use correct non-sticky splicing tape for jointing.

6 Tape stretching. All tapes should stand a steady pull of 4 lb without breaking, but the stretching yield may occur at half this value. The tension during normal recording or replay should not exceed about 6 oz. Greater tensions may occur during spooling or, in some cases, effects of moisture on tightly spooled tapes are said to cause stresses sufficient to stretch the tape. A reasonably good tape should not elongate more than about 0·15 per cent when a tension of $1\frac{1}{2}$ lb is applied for 1 minute.

7 "Print-through." When a recorded tape is stored on a spool undisturbed for a while, weak copies of the recording may be impressed magnetically on adjacent layers, causing pre- and post-echoes. The level of printing may be increased if any accidental stray AC magnetic fields happen to influence the tape (hence store away from mains units, TV scan systems etc). High temperatures, tight spooling and using thin (long-play) tapes are all factors tending to increase "print-through." Re-spooling tapes before playing will often reduce the magnitude of the echoes substantially, but a carefully applied partial-erase current applied during re-spooling has been found to remove the print-through without appreciably affecting the signal. This is said to be due to the fact that the printed echoes lie only on the extreme outer layers of the tape coating.

8 Noise. Ideally it is thought that the noise on a tape could be −70 dB on the signal, but in practice, owing to a variety of causes, it is unlikely to be better than −50 to −60 dB. Some of the principal causes are—

(*a*) Variation in particle size and distribution.
(*b*) Accidental DC magnetization of head or tape.

(c) "Modulation noise," due to oscillator or other circuit noise which appears when a signal is recorded.

The best recording studios now use special dynamic noise-suppressing "compandor" techniques in order to reduce low-level tape noise to the vanishing point (the Dolby A system). The Dolby B system is used on tapes, cassettes and cartridges.

Tape Table

Playing Time per Track

Type	Diameter of reel (inches)	Tape length (feet)	$1\frac{7}{8}$ in./sec (h m)	$3\frac{3}{4}$ in./sec (h m)	$7\frac{1}{2}$ in./sec (h m)
Standard Play	5	600	1 4	32	16
	$5\frac{3}{4}$	900	1 36	48	24
	7	1200	2 8	1 4	32
Long Play	$3\frac{1}{4}$	300	32	16	8
	5	900	1 36	48	24
	$5\frac{3}{4}$	1200	2 8	1 4	32
	7	1800	3 12	1 36	48
Double Play	$3\frac{1}{4}$	400	42	21	10
	4	600	1 4	32	16
	5	1200	2 8	1 4	32
	$5\frac{3}{4}$	1650	2 54	1 27	44
	7	2400	4 16	2 8	1 4
Triple Play	3	450	48	24	12
	$3\frac{1}{4}$	600	1 4	32	16
	4	900	1 36	48	24
	5	1800	3 12	1 36	48
	$5\frac{3}{4}$	2400	4 16	2 8	1 4
	7	3600	6 24	3 12	1 36
Quadruple Play	3	500	1 4	32	16
	$3\frac{1}{4}$	800	1 24	42	21
	4	1200	2 8	1 4	32

Tape-recorder decks and machines

There are quite a large variety of tape-recorders available, either in the form of decks for building-in or as complete portable units which may be "free-standing" in a suitable compartment of a Hi-Fi ensemble. The accompanying electronics may be fixed to the deck or may be arranged for adjacent mounting.

The cheapest commercial units available cannot reasonably be expected to give completely flutter-free music recording nor can they be expected to have an indefinite life if they are subjected to hard and fairly continuous use. At the other end of the price scale, the fully professional studio or instrumentation type of tape-recorder can be very costly and will normally be unnecessarily robust and elaborate for the average user. A few "replay-only" tape-units have been introduced, but in view of the fact that pre-recorded programme tapes are both expensive and limited in availability they have not yet found widespread use. A number of semi-professional tape-decks and cassette machines are available which are capable of giving very good musical reproduction.

Although they are less sensitive than gramophone reproducers to external shocks and vibration when running and do not normally require levelling, so far as the need for general cleanliness and correct mechanical treatment are concerned tape-machines call for the same order of care, if the best results are to be obtained. It might be thought that tape-machines are more "fool-proof" than disc machines, but in fact many things can occur to prevent the tape from being transported properly or from making proper and intimate contact with the heads. Some tapes may shed dust, and accumulations of this and other foreign matter on the tape-drive, -heads or -guides may cause considerable trouble. Perfection of mechanical alignment of all the essential parts such as heads and guides, and the accuracy of the "running fit" of the drive spindle bearings are of paramount importance. Expensive hand fitting on a mass-produced machine is not to be expected and although mass-produced machines can be brought up to quite high standards of performance if an expert with proper equipment is able to spend some time perfecting alignments, trueing shafts, bearings etc, this is obviously not a job for the amateur.

Cassette-loading tape-decks have been produced but, for the most part, domestic tape-decks use small cine-type spools. The larger professional machines use either the US NAB spools or the German "Magnetophon" original type of single-sided plate spool. All these types of spool are the subject of British and International standards so as to ensure interchangeability.

The best professional types of tape-machine usually have a three-motor drive system, the tape-transport capstan being driven by a synchronous type of motor and each spool being driven by a hysteresis or induction type of motor. A light braking action on the feed-spool motor and the right amount of driving torque on the take-up-spool motor enables correct tape tension to be maintained during recording, reproduction, re-winding etc, the correct conditions on the motors being established by the controls.

There are, however, quite satisfactory machines in which a single synchronous motor drives both the capstan and the spools, a "slipping clutch" and an idler or belt drive being provided to give the correct driving torque to the spools. This idea of a slipping-clutch friction drive has been used for many years on professional cine-projector spool drives in order to maintain constant film-spooling tension.

The usefulness of a domestic tape-recorder can be further extended by synchronizing it to home cine equipment, thereby adding a sound-track to otherwise silent home movies. Various methods of doing this are available, eg the Bell and Howell Filmsound system.

Some aspects of the care of magnetic tape and the various types available have been covered in the preceding section on the tape medium. It may be reiterated here that tapes should be carefully stored on spools in boxes in a cool place away from any stray magnetic fields. This means that a suitable compartment away from mains equipment etc should be provided for keeping tapes in when an integrated Hi-Fi system and its cabinet work are being planned.

The way in which signals are recorded on magnetic tape has also been introduced in the preceding section. The need to apply the correct high-frequency bias is of paramount importance in order to obtain the maximum sensitivity and the lowest distortion from the tape. It is also necessary to ensure that any earlier recording or noise is erased from the tape prior to making a recording. This is done by applying the full bias-oscillator output to the erase-head, which precedes the recording-head. A much higher signal level is applied for erasure than for biasing. Theoretically, the bias level should be accurately set for each different type of tape used. The optimum bias level is generally taken to be that which gives the maximum output for the lowest value of distortion. This usually occurs at a bias slightly in excess of the value giving maximum tape sensitivity, ie the value which gives the maximum signal output for a given input. There is sometimes a pre-set bias-level control provided at the back of tape-recorders. If an output-level meter is available the user can set the bias control slightly beyond the maximum output point. If he has an oscilloscope or more sensitive distortion-measuring equipment, he can set the bias in a more scientific manner. Fortunately, the

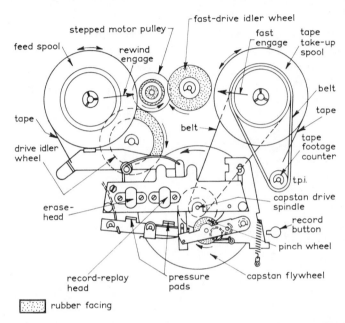

Fig 23. *A typical tape-drive traverse system used on a modern commercial tape-recorder. A synchronous motor drives the capstan flywheel via a rubber-tyred wheel in order to traverse the tape at a constant speed during recording or playback. The take-up spool and tape-footage counter are driven from the flywheel by flexible rubber belts. For fast re-spooling, a reverse idler is engaged, speed change being accomplished by engaging a larger-diameter step on the motor shaft. Slipping clutch arrangements are used to drive feed or take-up spools so as to maintain the correct tape tension.*

setting is not very critical in that the performance does not suffer very much if the control is fixed at one setting for different makes of tape, as is necessarily the case for most domestic tape-recorders.

Many professional tape-recorders have three separate heads. First, the erase-head, then the recording-head and finally the reproducing-head. The electrical and mechanical requirements for each type of head are different. The erase-head has to carry sufficient power from the bias oscillator to saturate the tape magnetically. It generally has a fairly wide gap (100 microns). The recording-head has to have a smaller but well-defined gap (10–20 microns) and need only carry the bias oscillator and recording-signal power. It must, however, be accurately made and aligned with its gap perpendicular to the tape length so as to give a good high-frequency response. The reproducing-head should have the finest possible gap (3–5 microns) and must also

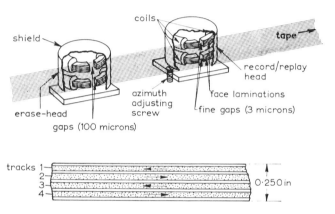

Fig 24. *A view of the head faces as seen by the tape in a modern four-track tape-recorder. The erase-head on the left has a wider gap and higher power windings. Recording and replay is accomplished by the other (dual-purpose) head, which has small very accurately defined gaps, sectionalized lamination and winding stacks for the separate tracks, and an overall magnetic shield which is cut away over the front face. Light pressure-pads are normally provided in order to keep the tape in close contact with the gap faces.*

be accurately aligned. It must generally be fitted with a magnetic shield so as to avoid hum pick-up. For multi-track recorders the heads must be suitably positioned and sub-divided so as to produce the correct tracks. The head faces which contact the tape must have a very high surface-finish, to ensure good tape-contact and low wear.

Most commercial recorders use an erase-head and a combined record/replay-head with a fine gap of 3–5 microns.

We have noted that magnetic tape-recorders are arranged to run so as to traverse the tape past the heads at one or more of the various standard speeds: 1⅞, 3¾, 7½ or 15 inches/sec. An approximate guide to the frequency response available with good, well adjusted heads and good-quality tape is that the response in kiloHertz is approximately equal to twice the speed in ips. Thus 1⅞ inches/sec gives a good intelligible speech recording with the maximum playing time, 7½ inches/sec gives Hi-Fi recordings, whilst 15 inches/sec is seldom used except for some professional recordings. The majority of domestic tape-recorders run at 3¾ and 7½ inches/sec. They are normally set to record two tracks—the heads being offset with regard to the centre of the tape, so that a second pair of tracks are brought into play by turning the spools over and changing them left to right, giving four tracks in all. Where double vertically-in-line record- and replay-heads are provided, two stereo tracks can be provided. Double offset heads will provide four-track recording. The use of only part of the ¼ in. wide tape is bound to have some effect on the output and hence reduces the signal/noise ratio, particularly for four-track recording. Four-track recording is, however, quite acceptable on a machine with good heads, using good tape, preferably a super-flexible long-playing polyester grade.[8]

Volume-level indicators
The position of the volume control must be set for any recording so that the average modulation level on the tape is high enough to give a good signal/noise ratio, but not so high as to give rise to distortion through overloading. Volume indicators are usually of the electronic "magic eye" type or are fast-acting pointer indicator meters. "Riding" the volume control to maintain optimum recording conditions can be quite difficult, particularly if wide variations in sound level occur at the microphone. Electronic volume-limiter circuits are incorporated on some recorders, so as to enable a high average recording level to be maintained without fear of overloading. Circuits of this sort have long been used in disc recording, where it is essential to avoid the risk of running one groove into another. The audio signal is taken through a side circuit where the alternating voltage is rectified and used to bias a diode in such a way that its impedance is reduced when large signals occur. The diode may form the lower limb of a potentiometer circuit in the main signal path. A fairly fast-acting time constant is required (a few milliseconds) so as to avoid passing any appreciable overload peaks. The release time must not,

however, be too short (preferably approaching $\frac{1}{2}$ second), or audible "thumps" may be produced by excessively rapid changes of gain. Another type of limiter uses a low-current lamp of very low thermal inertia closely encapsulated with a photo-sensitive diode. The lamp is fed from the side-circuit voltage, as before, with the advantage that the slight thermal inertia of the lamp reduces the risk of "thumps." The attack time is reduced, however.

EQUALIZATION

The natural recording and playback processes on magnetic tape and various inevitable losses do not result in a flat frequency response. The playback equalization required differs for the various recording speeds and these curves have been standardized except for some variations to allow for differing losses in some designs (Fig 25).

TAPE-RECORDER CONTROLS

These have to perform both electrical and mechanical functions and various interlocks are essential. The bias-and-erase oscillator

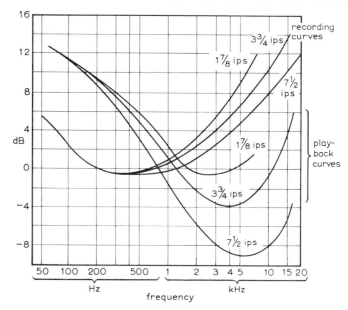

Fig 25. Tape-playback equalization curves which compensate for losses and for the natural rising response obtained when recording on magnetic tape.

must be cut off during rewind and playback, and the equalization must be changed to suit the speed engaged. The tape should preferably be withdrawn from the head faces, any pressure-pads being also withdrawn, during the rewinding operation.

Favourite methods of arranging controls are either by rotary switch, by a "joystick" type of control working in orthogonal slots or by "piano-key" types of push-button. The controls should be ergonomically sound in design, that is, they should come readily to hand and be easy to operate. They should also be clearly labelled and their various operations should be arranged in a logical sequence so that, for example, a recording is not likely to be accidentally erased by switching to the record position in mistake for the playback position.[9,10,22]

ELECTRICAL CONNECTIONS TO TAPE-RECORDERS

Separate input terminals or sockets are usually provided for radio input, gramophone pick-up inputs and microphone inputs. Where stereo recording is catered for double input sockets have to be provided in each case. Correct polarity or "phasing" must be preserved throughout the system in the case of stereo. Overall acoustic/electronic professional phase-checking apparatus is used in large studios, but the amateur must usually rely on simple phasing checks such as trying a centrally disposed soloist or talker to check that a central stereo-sound image results on reproduction. Wrong phasing polarity results in a vague, blurred image and, in some cases, a lack of low-frequency response, ie audible as reduced bass or cancellation of any residual amplifier hum.

Pick-up and microphone input terminals were at one time often high-impedance, suitable for piezoelectric crystal pick-ups and microphones. Low-impedance pick-ups and microphones require extra amplification and must be fed via suitable input transformers or into low-impedance terminals. High-input impedance levels are now usually 50 k-ohms to a few megohms and low impedances 30 ohms to about 200 ohms.

The advent of transistorized tape-recorder amplifiers operates against the use of high-input impedances on the score of noise. The optimum input impedance for many transistor circuits is 2 to 10 k-ohms and some are now provided with integral input transformers of this order of impedance.

If external input transformers are used to match low-impedance pick-ups or microphones to tape-recorder inputs, the external

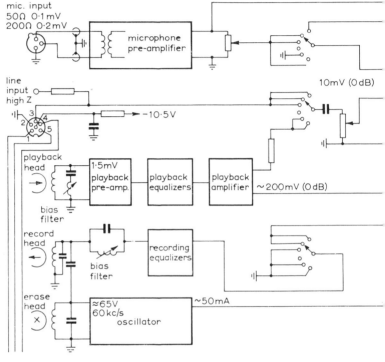

Fig 26. A typical tape-recorder functional schematic s

metal cases or laminations of the transformers should be earthed to the input cable shields of the microphone etc and to the recorder earth terminal, if one is provided. Needless to say, no contact must be made with the chassis of any AC/DC or transformerless equipment on account of the risk of electric shock in the event of the mains plug being reversed, so as to make the chassis live to earth. A Hi-Fi dealer's advice should be sought in cases of doubt, as isolating transformers may be needed.

The output connections of tape-recorders are usually low-impedance terminals for external loudspeakers etc. These are normally quite suitable for connecting to the radio-input terminals of most amplifiers. Some tape-recorders incorporate links or special plug arrangements to mute the internal loudspeaker; professional machines have line output terminals suitable for feeding other amplifiers. Fig 26 gives a block schematic showing

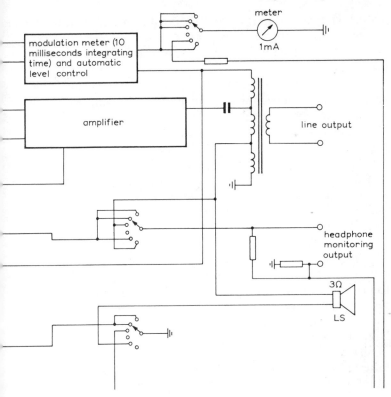

phone and other input circuits, extra outlets etc.

the functions of the various circuits incorporated in a portable professional recorder which gives full facilities.

Synchronization of tape-recorders and home cine equipment

The simplest type of synchronization consists in using stroboscopic bars either printed on the reverse side of the magnetic tape or on a pulley disc driven by the tape or capstan etc on the tape-recorder. The recording is made during the filming, using preferably a sound-and-vision start marker similar to the clapper board issued in film studios, thus enabling the starts of film and tape to be lined up for replay. The tape strobe bars must be illuminated by stray light from the cine-projector beam, diverted if necessary with a piece of glass or mirror in the beam. The usual cine shutter running at 16 frames/sec with a three-bladed shutter gives 48 light pulses/sec. Where the projector speed can be adjusted by a manual control it is fairly easy to bring it into synch with the tape strobes by making these appear stationary. Alternatively, the tape speed must be varied. This is not easy to do on most tape-machines.

There are quite a number of commercial systems available for amateur and professional use, which offer automatic synchronization of magnetic sound and the film projector. One system is the "loop synchronizer" where a loop of tape from the tape-recorder is passed through a system of guides and rollers mounted on, or near to, the projector. A spring-mounted jockey roller is sensitive to the absolute speed of the tape and its deflection adjusts the value of a variable resistor in the projector-motor circuit, thus making the system speed-regulated and synchronous. Another system is the "Synchrodek." This is a differential gearbox which has a drive from the tape loop and another from the projector motor or sprocket shaft. The speed-difference differential gear output again is made to control the projector motor voltage so as to maintain synchronism. Full details of the various synchronizing systems and equipments can be obtained from books on cine equipment and recording.[8,9] A complete system is the Bell and Howell Filmosound.

Reasonably priced TV video-tape-recorders are now becoming available and provision is made for recording a ready-synchronized sound-track to accompany pictures recorded from the camera.

Remote-control systems

Systems using coded ultrasonic waves are available from Saba and other makers for the armchair control of TV and Hi-Fi amplifier systems, including switching, volume- and tone-control adjustments etc. There is obviously scope for adapting such systems to the remote control of tape-recorded commentaries or other accompaniments to amateur film or slide projection.

Reel-to-reel stereo tape-recorders, cassette and cartridge machines

Two-track reel-to-reel tape-recording machines with separate vertically-aligned heads and their own reproducing channels were produced by EMI in the UK in the 1950s. Pre-recorded stereo programme tapes on spools were available from them and from various companies in the USA for many years. With the later development of stereophonic gramophone records which used cheaper and more generally available equipment, pre-recorded stereo reel-to-reel tapes did not achieve very wide use. However, magnetic tape is obviously a good medium for carrying stereo channels, as four or more separate tracks are readily accommodated. During the 1960s, compact and convenient tape-cassette containers and record/replay machines were developed. The alternative forms known as cartridges were also developed, and a large repertoire of recorded programmes is made commercially available on cassettes and cartridges by a number of manufacturers. The compact form and the narrow magnetic tracks required the use of slow transport speeds and small head sizes, with the result that the noise and the response were not initially up to high-quality sound requirements. Much intensive development in recent years has improved the quality which can be obtained to a very considerable extent. The best tape-cassettes and cartridges which incorporate all the improvements are now capable of the highest performance and are also offered with quadraphonic discrete four-channel pre-recorded programmes. There is little to choose between these top-quality cassette recorders and similar quality reel-to-reel tape-recorders, though the latter are probably easier for the amateur recordist to use and to maintain in top condition, in spite of the inconvenience of handling and storing the larger tape-reels.

If one wishes to use pre-recorded tape programmes, one is more and more bound to use the cassette or cartridge systems. The table on page 94 summarizes the relative advantages and disadvantages of the two systems. The use of the new high-coercivity tape materials, ferrite magnetic recording and reproducing heads with extremely fine and accurate gaps, precision tape-drive transport mechanisms and dynamic noise-suppressing circuits, in particular, have made the better grades of cassettes and machines a good proposition for high-quality home recording and reproduction. Cartridges are larger, use wider tape run at $3\frac{3}{4}$ ips and so may be less critical than cassettes. However, the endless loop characteristics of cartridges and the general lack of fast wind-on facilities have made cartridges more popular for continuous music and car players, rather than for general home use. It seems that, in terms of availability and improved quality, cassette machines are achieving a rapid growth rate.

Figs 27 and 28 show the tape and head layouts used for

Advantages of tape-cassettes

Cassettes are more compact than cartridges, about one quarter the volume.

Cassettes give longer playing time, as a rule, eg 120 mins compared to 90 mins.

More programmes are usually obtainable on cassettes as well as a wider selection of types of music.

Cassettes were designed from the start for recording as well as replay and standard precautions such as anti-erase stops are provided to avoid accidents.

Blank cassettes are easily obtainable.

Cassettes have low friction, are easily kept clean and may be dismantled for splicing, etc.

Cassette machines usually have good fast forward and rewind facilities, thus facilitating positioning and locating starts of items etc.

Advantages of tape-cartridges and some limitations

Cartridges potentially offer better high-frequency response, as the speed is $3\frac{3}{4}$ ips as compared to $1\frac{7}{8}$ ips.

Cartridges use an endless tape loop and will play continuously for background music etc.

Greater friction and wear is involved in the cartridge system. Lubricant is used in order to reduce dust accumulation and head-wear.

Blank cartridges and cartridge-recording as well as reproducing machines are not as easily available in Europe as cassettes.

Relatively few cartridge machines have a fast forward wind facility. When available, it is usually only about one-tenth of cassette fast-wind, due to endless loop friction.

Cartridges are not very suitable for editing and splicing, even when they can be dismantled.

Cartridges offer more track space for stereo and quadraphonic recording, owing to the extra tape-width available.

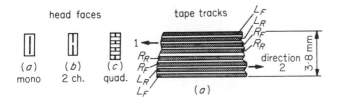

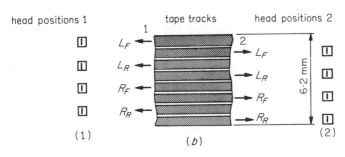

Fig 27(*a*). *Quadraphonic cassette tape and compatible heads;* (*b*) *Quadraphonic cartridge tape and heads arrangement.*

Fig 28. *The mechanism of the NEAL cassette-recorder. The tape heads, locating pins, the spool-drive bosses, the tape-drive capstan shaft and the pinch roller are visible* (NEAL, North East Audio Ltd, Newcastle).

cassettes and cartridges. The ease with which mono, two-channel stereo and quadraphonic compatibility is achieved with tape machines and cassettes is obvious due to the way in which deeper head gaps can scan several channels and produce the sum of the instantaneous signals in each (Fig 27).

It should be emphasized that in view of the complexity and the precision work involved in a top-class cassette machine, it is necessary to pay a price possibly between £100 and £200 to buy such a machine, possibly slightly more than an equivalent tape reel-to-reel machine. The extra investment, however, should ensure high quality.

Cassette machine details

There are now so many cassette tape-recording/reproducing machines available at a wide range of prices, that it is difficult for a non-expert to decide which machine to buy and what price to pay. It is apparent that really top-grade cassette machines are now becoming largely comparable to reel-to-reel machines in performance on a price-for-price basis, but compared to reels cassettes are much more compact, convenient and easy to store, either on shelves etc or in the many commercial storage boxes and cabinets now sold for accommodating cassettes and cartridges. The equivalent storage boxes for reels are much more bulky and expensive, some form of storage being essential so as to avoid tape spillage and damage. Bearing in mind also the rapidly growing number of high-quality pre-recorded cassette programmes, it seems that cassette machines are now of overwhelming interest to many home users. Reel-to-reel machines were dealt with earlier in this section and thus we will confine our description here to cassette machines.

Figure 29 shows the construction of a modern tape-cassette. It is worth buying good-quality cassettes, even if they are expensive. The quality of the tape as regards its response, freedom from abrasiveness, drop-outs and dust-shedding is of such importance alone to justify this slightly increased expenditure. In addition, good cassettes are much more free of defects such as sticking or jamming. Many cassettes can be taken apart for repair or tape-editing, and kits are available for doing this.

The layout of good stereo cassette machines, with their simple controls, volume indicator meters and the way the cassettes are

Fig 29. *A modern high-quality tape-cassette using chromium dioxide tape. Side tabs can be removed so as to inhibit machine erase mechanism and to prevent accidental loss of recorded programmes. Lugs are provided to switch in chromium dioxide equalization* (Philips Electrical Ltd).

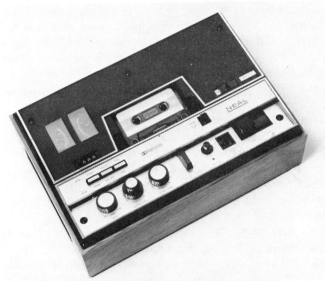

Fig 30. *A modern high-quality two-channel tape-cassette recorder which is fitted with Dolby B noise-reduction circuits. The main controls are ganged and two-volume indicator meters are fitted* (NEAL, North East Audio Ltd, Newcastle).

"posted" or inserted into the working recess are now familiar, and so we will not here describe the various makes of machines and the ingenious drive mechanisms used to transport the tape past the heads. These are extensively dealt with in the various audio magazines and periodicals. Fig 30 shows a modern high-quality cassette machine.

In addition to improved standard ferric-oxide tape, new types of tape in the form of chromium dioxide tapes and cobalt high-energy tapes are available in cassettes. The latter types can offer improved high-frequency response and lower noise, provided that the equalization and the bias levels are properly set on the machine for the tape used. This is most important if the full results are to be obtained. Some cassette tapes also have special antistatic and lubricating properties which reduce any tendency of tape layers to cling together and to cause wow and flutter.

As well as using high-quality tape, it is obviously very important to use a machine with properly aligned head gaps with regard to the 90° tape axis, as well as precision gap faces kept properly clean and free from tape dust. In fact, one should clean the tape heads, guides and capstan path on the machine almost as often as one cleans the stylus of a light-weight gramophone pick-up. Commercial cleaning kits are available, and some experts recommend the *sparing* use of isopropyl alcohol on cotton wool. If a machine is used regularly, it should be returned to the maker's recommended service agent about once a year for cleaning, head demagnetizing, checking and overhaul. In view of the necessarily complicated and delicate mechanism of cassette machines, this is obviously sound practice.

As an example of the performance now claimed for advanced heads, one can examine Akai's results which claim an indefinite head life and a response to over 17 kHz on chromium dioxide tape, using a high-precision single-crystal ferrite head set in a fused glass surround and gap filler. Total wow and flutter figures of 0·1 to 0·2 per cent on the best machines are undetectable on any normal listening.

It is, however, advisable to monitor the recording level carefully on the meters provided on most top-grade machines, so as to avoid overloading the tape with an excessive modulation level and thus causing distortion. As this means erring on the lower side, if anything, it is obviously important to use a dynamic noise suppression system, such as the Dolby B, in order to obtain a good signal-to-noise ratio (of 50 dB or better). Some machines use fast-acting peak-reading light-emitting semiconductor diodes to indicate instantaneous overload conditions.

Such results are indeed good, in view of the great convenience of the cassette system and it is worth taking trouble to see that they are maintained. Apart from cleanliness, it is worth ensuring that such faults as leaving the rubber-faced pinch wheel engaged in contact with the capstan shaft, causing flats and wow, are not possible if the machine is left with the play-button pressed on switch-off. Many machines have a solenoid-operated drive disconnection system in order to avoid this hazard.

The various noise-reducing systems which are so important on cassettes are the subject of much confusion to the non-expert, and thus the main types of these systems are outlined in the next section.

Performance specification for a high quality tape-cassette machine

Tape transport
   Type: 3M Wollensak heavy-duty mechanism, featuring large capstan and flywheel assembly, self-aligning pressure roller, and cassette guard.
   Tape speed: 1·875 inches per second (4·76 cm/second).
   Fast wind and rewind times: Less than 45 seconds for C60 cassette.
   Wow and flutter: Less than 0·15 per cent r.m.s. DIN weighted.
   Cassette protection: Fully automatic complete disengagement of mechanism at end of tape or in the event of a stalled defective cassette.

Noise reduction
   NR In/Out: Operates Dolby B noise-reduction circuits by solid-state switching.
   Stereo/Mono: Enables recording on both tracks from a mono input.
   CrO/Reg.: Selects different bias levels, recording pre-emphasis and replay networks for either chromium dioxide or low-noise tape by solid-state switching.

Meters
   Type: Twin peak programme meters reading positive and negative signal peaks. Meters indicate true pre-emphasized recording signal and equalized playback signal.
   Electronic time constants are 2 ms rise time and 200 ms fall time.
   Calibration: 0 dB corresponds to 22 mM/mm flux density in gap.

Recording amplifiers
   Distortion: Less than 0·1 per cent from any input to head at 0 dB level.

Input amplifiers:	*Input for 0 dB recording level*	*Input impedance*	*Overload margin*
Microphone	250 μV	2·5 k-ohms	better than 40 dB
High-level line	80 mV	2 M-ohms	better than 40 dB
Low-level line	2·25 mV	5 k-ohms	better than 40 dB

Replay amplifiers

Distortion: Less than 0·1 per cent from head to all outputs at 0 dB level.

Frequency characteristic: Chromium dioxide, 70/3180 μs; Ferric oxide, 120/3180 μs.

Outputs: Monitor output, 110 mV fixed from an output impedance of 4·7 k-ohms. Line output (DIN), 2 V controlled—output impedance 1 k-ohms. Headphone output, 2 V controlled—output impedance 50 ohms. Output levels referred to 22 mM/mm flux density. All outputs protected against short-circuit, without distortion of any output.

Frequency response

Ferric oxide tape, 35 Hz–12,000 Hz (+1 dB–3 dB). Chromium dioxide tape, 35 Hz–15,000 Hz (+1 dB–3dB).

Crosstalk

Record on left (right), replay on right (left), better than 45 dB.

Signal-to-noise ratio

Unweighted signal-to-noise ratios referred to a flux density of 25 mM/mm measured over a bandwidth of 5 Hz to 20 kHz.

Chromium dioxide tape, Dolby off, 55 dB. Low-noise ferric tape, Dolby off, 53 dB. Chromium dioxide tape, Dolby on, 57 dB. Low-noise ferric tape, Dolby on, 56 dB. (Based on data provided by NEAL, North East Audio Ltd, Newcastle.)

Dynamic noise-suppression systems

It was realized long ago that the quality and intelligibility of many systems is improved if a continuously varied noise-suppressing system of some kind is applied. It is apparent that unwanted background noises are completely masked when the system is carrying loud and continuous sounds, but hiss etc may be very disturbing during quiet or silent passages. Fixed filters or tone controls have the disadvantage of reducing the quality of all sounds, loud or soft. Dynamic systems are often in the form of "compandors" or fast-acting side-circuits which almost instantaneously change the gain of the main programme channel before the point at which interference is generated. Thus soft sounds are made louder relative to the background, but the overall volume range of the programme is reduced. Unfortunately, the operation of these limiters or compressors is liable to introduce audibly disturbing noises unless they are very carefully designed. The disturbances have been likened to thumps, surges or "breathing" sounds, depending on the speed of action etc.

In the course of research some 10 years ago, Ray Dolby realized that an important advance could be made by exploiting the fact that the masking effect makes it much less likely that disturbing sounds can be detected as long as they are close in frequency to the spectrum of the signal which they control. There was an urgent requirement to improve the signal-to-noise ratio of master tape-recordings, as the tape hiss exceeded that added by subsequent disc recording and playback. The studio Dolby A system works by splitting the signal into four bands covering the audio range, and raising the level of low-level components in a controlled manner, leaving high levels unaltered. The volume range is automatically restored on replay. A simplified Dolby B system has been evolved for tape-cassette machines and similar home systems, where the signal-to-noise ratio is deteriorated by hiss. The basic principle is illustrated in Fig 31.

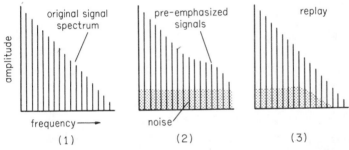

Fig 31. The action of a pre-emphasizing noise-reducing system of the Dolby B type. (1) The frequency spectrum of a typical programme with lower amplitudes at the higher frequencies. (2) The higher frequencies boosted in level prior to the recording or other process at which noise or hiss is introduced. (3) A reduction of level of all frequencies in the pre-emphasized part of the spectrum restores the original frequency response and reduces the noise at the same time (Dolby Laboratories).

Dolby noise-reducing system details

The Dolby system uses similar circuits on recording, prior to the point where interference is met, and on replay. No separate control signals are transmitted and thus the system is eminently suited for use at points separated in time and place, such as a tape recorded in one place and subsequently replayed elsewhere on another machine, a processor "black box" merely being inserted at the appropriate point in the circuits. High-level signals are passed straight through an amplifier, which also has signal-adding or subtracting facilities. A side "differential network" operates on lower-level signals only, in four frequency bands: 20–80, 80–3000, 3000–9000, 9000–30,000 Hz in the A system. Normally, any given band signals 40 dB or more below peak levels are boosted in level with an attack time suited to the signal frequency and rise time, adding the result to the main recorded signal. On replay, similar filters in the side-circuit detect the boosted low-level signals in the various bands and instantaneously subtract these from the composite main signal, thus reproducing the original programme dynamic range, with an effective reduction of the noise and hiss level of 10 to 15 dB or more, sufficient to make the interference inaudible in most cases. The use of the four bands above means that hum and middle-frequency noise are reduced as well as hiss. A further advantage is that any residual background noise, left after Dolby-processed replay, is constant, not varying as in many compressor systems, which is more distracting.

The simpler Dolby B system applies to cassette recorders, FM broadcasting and other more widely used systems where the addition of hiss is the major noise problem. A high-pass filter operating broadly above 2000 Hz is used in the side system, boosting low levels in a generally optimum fashion which is adapted to the programme. The Dolby B decoder is simply added in series with the home reproducer when a Dolby B processed programme is played. The low levels are automatically restored, as before, in the amplifier-subtractor stage. Unlike the A system, which requires an accurately complementary processor on re-production, the B system is substantially compatible, in that if B processed tapes or broadcasts are played without a B processor on reproduction, there is an increase in higher frequencies at lower levels. These can be reduced by turning down the top tone control slightly, at the expense of some loss in the high frequencies at higher levels. One can find an optimum setting by trading some top loss against noise reduction in quiet passages. However, more and more good-grade tape machines are being fitted with the Dolby B system, which can be switched in as required.

The Philips Dynamic Noise Limiter (DNL)

A number of variable noise-suppressing systems have been evolved. The Philips system is applied to reproduction only, and thus the question of compatibility does not arise as it can be applied to any desired programme. It also allows high-level signals to pass through unaltered, but high frequencies are cut when only low levels are present, this being justified by the fact that high-frequency components are not greatly missed at low sound levels, but that the hiss level is much reduced. The result is a great improvement on a fixed top cut tone control.

The incoming signal is split into two parallel paths which are recombined in an adder circuit. One path is an all-pass filter stage, which has a flat frequency response but which introduces a 180° phase change. The other path has a high-pass filter which only admits frequencies above about 4 kHz to a variable attenuator whose loss can be instantaneously controlled by a level-sensing detector operated from the input. The presence of high-level signals effectively blocks this path, whilst at lower input levels, considerable amounts of the frequency spectrum above 4 kHz are passed on to the adder, where they cancel out the identical high-frequency signals from the input, which are now 180° phase changed in the lower path. The result is to reduce the high-frequency components in the output, reducing the hiss at the same time, by amounts which may be between 10 and 12 dB.

The system is simple and ingenious, as it is fast-acting yet virtually free from distortion or surges etc. It cannot be quite as effective as a pre-emphasizing type of system such as the Dolby, when noise is introduced at a previous specific stage, such as in a recording. The Philips system has been found very effective for tape-cassettes and is widely used on these machines.

5

The use of microphones

The microphone function

Basically the function of a microphone is to pass on a correctly processed electrical copy of the sound field at a point of interest. This may be in the near sound field or in the distant sound field of a source. A microphone does not respond in the same way as the binaural human hearing system and thus its response to a given type of sound field depends on the microphone characteristics which, in turn, dictate its placement.

In general, microphones have to be placed fairly near to sound sources in order to obtain a signal having a correct "acoustic perspective" with regard to discrimination against unwanted reverberation, background sounds and other air disturbances having audible or surge-producing components. Relatively distant microphones may be used for speech and music in studios and other fairly large and quiet enclosures but a close microphone technique is necessary for speech pick-up in the majority of cases.

The performance of distant microphones may be optimized for plane-wave incidence but close-talking microphones have to work under the complicated acoustic conditions prevailing around the human mouth and the secondary speech sources such as the nostrils, throat and chest surfaces etc. Human speech is normally heard at a distance of 1 metre or more where the wavefronts are more nearly plane. An ideal close-talking microphone should be designed to reproduce natural-sounding speech on either a loudspeaker or an earphone. In the latter case, the head-diffraction effect is absent and it is usually considered that a telephone microphone should have a rising response above 1 kHz to compensate for this.

Present two-channel stereophonic reproduction produces an

acoustical illusion of source direction mainly by an amplitude-to-time difference conversion which occurs at the ears when two spaced loudspeakers are fed with different amplitudes obtained from crossed directional microphones. If spaced microphones are used, an inter-channel time-difference element is introduced, but this is normally considered to be of secondary importance.

Hence we see that it is not easy to state the requirements of a microphone in simple terms and a considerable amount of subjective experiment has been involved in arriving at the specifications for given types of microphone. The extent to which ideal results are approached is limited by the overall state of the art and the potential performance obtainable from existing microphones.

Thus there is a fair amount of scope for experiment in microphone positioning and subsequent electrical equalization or tone control in any given set-up.

Types of microphone
The operating principles and the general forms of construction of most present types of microphone have been used for many decades and it is noteworthy that no widely applicable new principles have emerged. The tendency is to improve the performance and the reliability of known types by better design and the application of new materials.

Thus the carbon microphone is still almost universally used for telephones, as the large amount of research which has been applied over the years to the refinement of the carbon granules, the shape and material of the electrodes and other aspects of the design has improved performance and reliability to a very great extent. The carbon microphone also has useful subsidiary amplitude transfer characteristics such as lower-level noise suppression due to origin distortion, and high-amplitude clipping. If a linear type of microphone is substituted for a carbon microphone, amplification is needed and the lower-noise gating and upper-level clipping characteristics may have to be simulated in noisy surroundings.

Other forms of microphone transducer include silicon "strain-gauge" types and piezo-transistor types. Generally they have not approached the carbon microphone for efficiency. However, on account of its basic limitations as regards noise, linearity and inability to withstand certain types of vibration, the carbon microphone is not much used outside normal telephone applications.

Linear microphone transducers include moving iron, moving coil, piezoelectric and capacitor types. Efficient microphones may be made on all these principles but the moving iron is eliminated for use in high-quality wide-frequency range microphones for broadcasting, recording, acoustic measurements etc. All these types require amplifiers to raise their output level and in some cases to perform an impedance transformation.

In all cases either basically omnidirectional pressure types or pressure-gradient directional types may be produced. In the latter case the sound pressure has access to the front and to the rear side of the diaphragm, which is moved by the sound-pressure difference between the two sides at any given instant of time.

Present constructional trends and transistor microphone amplifiers

We see that microphone transducers may be basically divided into those in which the diaphragm drives the transducing element at a discrete point or area such as the apex and those in which the transduction mechanism is uniformly distributed over most of the diaphragm surface and is intimately associated with it. Such devices include ribbon microphones, capacitor microphones and piezoelectric plate microphones, in which a thin piezoelectric ceramic material is attached to the surface of a flat diaphragm. The potential gain in simplicity, reduction of mass, stability and robustness offered by distributed systems is apparent. They may also be more suited to the automated or integrated assembly methods which will be increasingly used in the future.

The availability of robust low-noise transistorized microphone amplifiers of very small size, which may in some cases be mounted in or near the microphone, has given more freedom to obtain an optimum compromise in the design of microphones. For instance, it may not be necessary to place so much emphasis on sensitivity, this being exchanged for robustness or a reduction in size. Head amplifiers may be powered over the microphone leads in various ways.

An important factor is always the signal-to-noise ratio obtainable at the input stage. When a microphone is associated with its own amplifier it is possible to obtain an optimum design compromise if a given minimum S/N (signal/noise) ratio is desired. The study of microphone and transistor noise performance is thus worth analysing in some detail. Special microphone amplifiers and mixer units are often needed if several microphones are used. These are expensive to buy but may be made up to published designs (see Fig 33, page 114).

Noise considerations at transistor input stages

Transistors generate several types of electrical noise but have an advantage over valves in that they are substantially free from microphonicity and heater-induced hum. The resistive component of the base input impedance gives rise to thermal agitation ("white") noise and the carrier conduction or gate leakage effects give rise to shot noise, whilst other noise sources are due to recombination and surface effects, including leakage. The general effect is that one can consider the output noise as being due to several generators in series, one producing a "red" 1/f noise spectrum, and others a "white" noise spectrum. Any shunt capacity on the input circuit, such as that of a piezoelectric or capacitor microphone will again modify the input circuit noise to a "red" spectrum. The overall noise varies considerably with the type of transistor and also the transition point on the frequency scale where the higher low-frequency red noise falls into the constant white noise level.

The noise factor of a bipolar or a field-effect transistor is the dB ratio of the actual noise from the device to the thermal noise produced in an equivalent resistor, referred to the input. The figure obtained depends on the centre frequency and band-width concerned. Audio devices are usually quoted for 1 kHz and a band of 10 kHz and may have noise figures between about 0·5 dB in the best cases to 12 dB or more. The S/N (signal/noise) ratio may be optimized for a given device in given circuit conditions by stepping up the microphone source impedance to a particular value, commonly between 1 k-ohms and 10 k-ohms. Operating the transistor at a fairly low power dissipation may help to minimize noise but there may be little improvement below a certain point and there may be a risk of overloading on large signal inputs.

It is seen that a microphone or other low-output transducer must be made up to the correct impedance so as to match the input transistor circuit to obtain the best signal-to-noise ratio. Ribbon and moving coil microphones are low-impedance devices owing to the sizes of wire or foil which must be used, whilst capacitor or crystal (piezoelectric) microphones represent high impedance reactive sources. Input transformers are thus often used for dynamic microphones, whilst capacitor types require impedance-transforming head amplifiers, commonly using low-noise field-effect transistors which can operate effectively at very high-input impedances.

Microphone properties and applications

Desirable qualities in a microphone are: small size, robustness, high sensitivity to desired signals with rejection of unwanted sounds (such as background noise, hum pick-up etc), uniform response to all frequencies, and suitable directional properties. No single microphone can meet all these requirements and the best results depend on the choice of microphone to suit the particular applications which the user has in view.

Certain classifications will occur to the user at once, for example, indoor and outdoor microphones, directional and omnidirectional microphones and microphones for close or distant talking. Detailed consideration of these categories will help to make clear the functional and constructional differences between the various types of microphones available. The distinction between high-quality and lower-quality microphones must be considered. Where cost must be taken into account, it should be remembered that a good microphone does not usually add much to the overall cost of an installation; a cheap one is thus a false economy. Good modern microphones are precision-made and utilize the best materials and manufacturing techniques. Each microphone is usually individually tested over its whole range of performance to a very exacting specification.

Directional properties of microphones

For many purposes a microphone needs to be sensitive to sound sources irrespective of the angle of sound incidence, as, for example, when the instrument is located centrally with respect to a group of performers. Such a microphone has been called non-directional since it has no favoured direction of acceptance, but "omnidirectional" is a better and more positive term to indicate that the microphone accepts sounds equally from all directions. It is well known that in air, as in water, the static pressure is the same in all directions; ie air pressure is naturally omnidirectional. In general, therefore, omnidirectional microphones are pressure-operated since they respond to changes in air pressure produced by sound waves.

Most early microphones were of the pressure-operated type, but later microphones were made which respond to changes in pressure gradient rather than to changes in pressure. In simple terms, if pressure is likened to the height of a hill, pressure gradient corresponds to the steepness of the sides of the hill and is thus a vector (ie directed) quantity. As air pressure varies with a signal, so does pressure gradient; thus an electrical device sensitive to pressure gradient changes can act as a microphone. Such a microphone is directional since, in its basic form, it is most sensitive to sound arriving from the front and back and least sensitive to sounds from the sides, top and bottom. The "figure-of-eight" directional pattern thus produced has certain advantages as it is sometimes possible to arrange for unwanted sound sources to be on the insensitive axis of the microphone. An even more useful sensitivity pattern is the cardioid (heart-shaped) polar curve. Cardioid microphones have the same discrimination between direct and random indirect sound as the bidirectional (figure-of-eight) types but they have the additional advantage of being insensitive to sound arriving from the back. This property is particularly advantageous for use on a stage, since the cardioid microphone helps to suppress unwanted noises coming from the orchestra pit or the audience and reduces echo effects from the back of an auditorium.

The growing popularity of stereo reproduction enhances the importance of directional qualities, since most stereo systems depend on the use of matched and accurately orientated directional microphones. Coincident "crossed axis" or spaced-apart microphones or a combination of the two types of mounting are used for stereo. It is worth experimenting with different angles and spacing so as to get the best stereo result for a given recording.

Close and distant talking

In a studio, particularly when music is being reproduced, the sound source is some distance from the microphone, whereas in public address systems, often used in a noisy environment, the talker is much nearer to the instrument. Some vocalists sing within a few inches of the microphone and obtain special effects by so doing. Finally, the commentator who has to provide a commentary during a live programme must put his lips close to the instrument.

These different requirements lead to the need for microphones with special characteristics. The single cardioid or bidirectional microphone gives a useful degree of exclusion of unwanted sound without the disadvantages of considerable size or complication. Microphones which operate on the pressure-gradient principle have another special property not found with pressure-operated types. With the former, the response to low-frequency sounds rises more rapidly as one approaches the instrument than do the middle- or high-frequency sounds. Ribbon microphones, which generally work on the gradient principle, are therefore well suited to studio use, but give low-frequency boost if used for close talking. This property can, however, be put to good use by introducing elements into the construction of the instrument which attenuate the lower frequencies so as to give an overall flat response when used for close talking at a prescribed distance.

Microphones with these characteristics can be designed so as to be eminently suitable for a commentator's use. Room or other noises, or the programme through which the commentator has to talk, besides being relatively distant, are robbed of most of their middle- and low-frequency content. The result is an effective suppression of everything but the commentator's voice. A microphone of this type is often fitted with a mouth-guard so that the distance from the speaker's mouth is accurately fixed. An alternative is the use of a microphone suspended on a neck halter. Small moving-coil types are best for this.

All microphones operate on the alternating components of air pressure of which the sound waves consist, so that puffs of air from mouth or nose and relatively large bursts of pressure which accompany certain consonants (such as "p" or "b"), can produce disastrous results when amplified from a microphone. One of the difficult problems of close-talking microphone design is to neutralize this "blasting" or "popping" without causing deterioration of the response. The commentator's microphone, as used by the broadcasting and television organizations, is so constructed that this unwanted characteristic is largely avoided and distortion-free speech of broadcasting quality is obtained. If this trouble is experienced when using a microphone close to the mouth, a simple windshield cut from porous foam plastic or a nylon stocking may be made, and this may also be used outdoors to protect against the noise of the wind.

Outdoor use of microphones

Some microphones, such as ribbon microphones, perform ex-
cellently in the calm air of a room or studio but are unsuitable for
outdoor use. In general a microphone to be used outdoors should
have smooth contours with no sharp edges to encourage air
turbulence or edge tones, and a robust moving element. Pressure
types are more suitable than gradient types, since any low-
frequency turbulence produced by the action of wind on a gradient
microphone is likely to be emphasized by the bass boost effect.

It is nearly always necessary to provide a microphone that is to
be used out of doors with a windshield. At its simplest, this may
take the form of a "glove" of foam material, sometimes silicone-
treated to repel water. For use in high winds, however, the microphone must be enclosed in a cage closely covered with a suitable
material, such as fine-woven wire mesh (about 120 meshes to the
linear inch). Windshields act by attenuating the velocity of the
wind air-stream, so as to prevent the setting up of serious turbulence
round the body of the microphone and, if carefully designed,
shields give little attenuation of the much smaller amplitude waves
of sound. Because the shield itself generates some turbulence, it
should be 2 or 3 in. in diameter to be efficient in very windy
conditions, so that this turbulence is kept as far away as possible
from the operating elements. This is specially important for
gradient microphones, which have an increased response for
low-frequency turbulent sounds generated near the sensitive
element. The simpler and smaller foam shields are still effective at
low wind-speeds and are also very useful to suppress noises due
to a speaker's breathing and explosive consonants or "pop"
noises.

Fig 32 shows a number of representative modern microphones
suitable for high-quality amateur use.

Impedance of microphones

Microphones are normally either of low impedance, as in the case
of moving-coil or ribbon units, or of high impedance, as in the
case of piezoelectric crystal or capacitor (condenser) microphones. Tape-recorders and preamplifiers may therefore have
microphone input terminals or sockets for either high or low
impedance. If it is desired to connect a low-impedance microphone
to a high-impedance input socket, an external microphone transformer must usually be used in order to obtain an adequate

(a) *An omnidirectional moving coil microphone suitable for speech and general purpose recording.*

(c) *A high-quality ribbon microphone for indoor recording of high-quality speech and music. The accurate bidirectional polar curve helps to eliminate unwanted background sound. Two of these microphones may be mounted close together at right angles for "crossed microphone" stereo recording.*

(b) *A compact unidirectional ribbon microphone with a polar pick-up characteristic which reduces unwanted background noise and reverberation.*

Fig 32. *A group of typical microphones.*

signal from the microphone. Small shielded microphone trans-formers are available with various impedance ratios.

Low-impedance microphones are usually nominally 30 to 50 ohms, but a medium impedance of about 200 ohms is now becoming popular as well, in order to get a better match to transistor input stages. High-impedance microphones such as crystal units may in fact have to be used with a step-down trans-former to match transistor inputs. Generally speaking, microphone impedance matching is not very critical, provided that sufficient signal can be obtained and that the frequency response is not affected. In cases of doubt the manufacturers concerned or the local dealers may be consulted.

Further references should be consulted for details of the design and construction of the many types of microphone available. [2,14]

Microphone placement

This is the art of placing the microphone relative to the performers so that a good "balance" is obtained between the different artistes, instruments etc, and the correct ratio between the direct and reflected (reverberant) sound as picked up by the microphone or microphones used. Generally, a close microphone position will give good "presence" and will eliminate excessive reverberation and the stray ambient background noises almost invariably present to some degree. The result may then be too "dry" or lacking in reverberant "colour," in which case use may be made of an artificial reverberation spring or plate equipment in order to add a controlled amount of reverberation to the final recorded result. Some of these devices are now within the range of the amateur's budget.

The reader is referred to books covering studio techniques and sound recording for more detailed treatment of microphone placement and studio recording techniques. [13,14]

Quadraphonic microphone arrays

Four-channel stereo experiments may be made by mounting four unidirectional cardioid microphones closely one above the other facing in directions mutually at right angles. Alternatively, four ribbon microphones may be used with a wad of acoustic damping material placed closely behind each ribbon, to suppress the back response as far as possible. Four-track tape- or cassette-recorders may be used (see Fig 27).

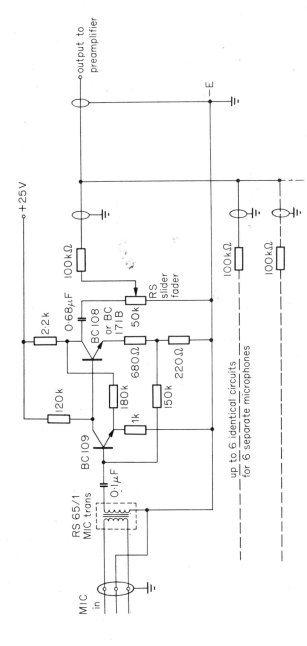

Fig 33. *A simple microphone mixer circuit to take up to 6 microphones of the moving-coil or ribbon type.*

6

Amplifiers and control systems

General properties of amplifiers

The old conception of an ideal amplifier as a device which merely increases the power in a circuit without modifying the signal in any other way has long since been superseded. Although it is still a prime aim to preserve the signal without distortion, modern amplifiers also perform quite a number of other functions. These can be broadly summarized as switching and control functions or filtering and equalization functions, together with a few useful subsidiary effects such as electrical damping of the loudspeaker. A good amplifier will also often provide smoothed and regulated power-supply outlets for running radio tuners, preamplifiers etc.[14]

Till fairly recently, the term amplifier mainly applied to a separate single-channel power-amplifier, fed from a radio tuner or a single-channel pre-amplifier. The advent of stereo led to the use of two electrically-separate amplifiers on one chassis sharing a common power-pack. A present trend is now towards single combined or "integrated" amplifier units which include a stereo tuner, control-amplifiers and power supplies. The enormous reduction in size, weight and heat dissipation resulting from transistorization and, in particular, the growing use of integrated circuits has made this type of combined equipment feasible and an economic proposition. Apart from a certain amount of metal required as a "heat sink" for output power transistors, most of the bulk of the new generation of amplifiers is in the power supply and in the control, equalization and switching functions. This overall concept greatly facilitates the design of compact, well styled and ergonomically satisfactory high-fidelity equipment which is easily placed in a living room and fits in well with modern furnishing arrangements. The elimination of the bulk, weight and large heat dissipation inevitably associated with valve-type power-amplifiers is particularly advantageous.

It was some time before all the difficulties associated with transistor circuits were eliminated and it could be said that the performance and reliability equalled that of the best valve amplifiers. The best modern transistor amplifiers embody a number of ingenious circuits and design features.

It is probably still true to say, however, that it is easier for the amateur to design and construct a good valve amplifier than an equivalent transistorized amplifier. Some of the specific difficulties which are encountered with transistors will be dealt with in the following pages. These include the inherently greater "origin-distortion" type of non-linearity displayed by transistors as compared to valves and the means available to minimize the resultant "cross-cover" distortion in push-pull power output stages. Another phenomenon which has to be guarded against is the tendency in transistors towards "thermal-runaway" or overheating, as the result of which a transient overload or output short-circuit may cause transistors to fail much more readily than valves. The advent of the more robust silicon types of transistors and the use of special circuits have greatly helped towards the elimination of these troubles.

Amplifier input and control functions

A well-designed amplifier must provide means for selecting, controlling and equalizing signals from any desired source. These normally include standard types of gramophone pick-ups, microphones, tuners or tape-recorder replay outlets. Mono and stereo switching arrangements may also be needed.

Unlike most signals, which can be assumed to originate with a nominally flat frequency response, the majority of modern high-quality magnetic gramophone pick-ups generate an output with a response related to the "constant amplitude" disc recording characteristic, except in so far as this is modified by their own internal mechanism. Modern moving-magnet, moving-coil and variable reluctance pick-ups give an output substantially proportional to frequency for discs recorded strictly according to this characteristic, which is now a generally accepted international standard (BSI, RIAA etc). The input stages of amplifiers are thus generally designed to compensate for this curve when switched to normal gram inputs. However, certain pick-ups, particularly piezoelectric types, are more conveniently designed so as to give a natural compensation for the standard recording characteristic and thus require a near-flat input amplifier position. Also a number of recordings have been made with different frequency responses, particularly those made before the introduction of the international standard. Again some records have unusual characteristics due to intentional special effects or unusual studio acoustics etc.

In other cases, some automatic limitation of the levels of low frequencies or high frequencies may have been imposed in order to avoid groove-wall breakdown or excessively deep or shallow cutting in stereo grooves or, again, to avoid too high a value acceleration and consequent tracing difficulties with the pick-up stylus. Microphone and voice characteristics may also have to be compensated. Thus a fair degree of independent bass and top "curve-bending" type of tone control is desirable. Circuits which do this are described later, including the well-known Baxandall tone controls. It is also desirable to eliminate any extreme low-frequency rumbles and hum and to reduce any appreciable high-frequency electronic or recording noise; suitable stepped filter characteristics with varying degrees of cut-off slope are often provided for both low-frequency and high-frequency cut-off. These filters have usually been designed as passive (non-amplifying) four-terminal networks in the past, based on conventional inductance and capacity filter network design theory. These components tend to be rather bulky and expensive by modern "integrated" standards, particularly in regard to the case of sharp cut-off low-frequency high-pass filters designed to remove rumble. Active filters designed to form part of an amplifier circuit are now gaining favour. These may often be designed as resistance-capacity networks which may be placed in the feedback paths of integrated or other transistor amplifier circuits (Figs 38 and 39, pages 132, 133).

As well as allowing the connection of various inputs at the correct impedance, the amplifier input stages have to meet extremely critical requirements as regards the range of voltage accepted. The two bounding conditions are the electronic noise-level limitation on the lowest signal levels that can be accepted and the overload capability limitation on the largest input allowed. These two requirements are not always compatible, in that a minimum noise figure is normally obtained when a transistor or valve is run at a low current, which may involve too low an overload ceiling.

The possible dynamic range on modern recordings is now considerably greater than that of a few years ago, owing to the fact that more sophisticated types of control, such as the "Dynagroove" system, help to offset the groove-tracing distortion limitations and also avoid the risk of excessive groove amplitudes and acceleration values by means of advanced peak-limiting devices. Dynamic-noise suppressors, such as the Dolby noise suppressor, mean that lower minimum signal levels can be accepted. The better modern gramophone pick-ups are also able to track higher recorded levels. It used to be assumed that the input stages of an amplifier could accept +20 dB above the nominal level at any input terminal. This figure now needs to be regarded as an absolute minimum and preferably should be considerably increased. If 30 dB is taken as a maximum requirement, then a high-output crystal pick-up with a nominal average output of 0·5 volt could give no less than 15 volts at peak levels. This is obviously in excess of the possible handling capacity of a transistor stage with a 12-volt supply and thus the input would have to be attenuated ahead of the first transistor. In other cases, the first transistor stage may have a considerable gain, possibly over 100 times as a voltage ratio. Thus an input level attenuated to 0·15-volt peak, which might be satisfactory for the input transistor, could overload the next stage with the output of 15 volts. This trouble might be reduced to negligible proportions by the use of suitable negative feedback over the two stages, and also by the use of 40-volts silicon transistors in the early stages. Thus the design of the input stages prior to the volume control point must be considered as a whole, and in fact this is now normally taken care of by the use of various forms of negative feedback circuit in modern designs. There is obviously considerable design flexibility as long as separate (discrete) components are used. However, some of the advantages of integrated circuits may be obtained by using commercial linear integrated transistor amplifier units and incorporating a volume-control in the external feedback circuit loop allowed for by the manufacturer. A volume control right at the input is not desirable on low-level circuits because, even when set to maximum, it may introduce enough loss to worsen the signal-to-noise ratio on a weak signal.

A number of special complete audio-amplifier integrated circuits are now made by leading manufacturers, who also supply details of the external components which have to be added in order to make preamplifier and output stages with tone control and volume control facilities etc (Motorola, National Semi-conductors).

Power-amplifiers

Many of the important properties and requirements of power-amplifiers have been well established over the years for valve amplifiers. Thus we are familiar with concepts such as matching to loudspeakers for optimum power transfer and proper damping of the loudspeaker bass resonance. Other factors are the power-handling capacity over the audio-frequency range, the amount and type of non-linear distortion produced, the response to normal and abnormal transients and surges. Other properties of push-pull valve output stages, in either class A or class A-B, concern normal centre-tapped output transformers or transformers with "ultra-linear" primary circuit taps for tetrode screen connections, tertiary feedback windings, tapped or series/parallel secondary connections etc.

Transistor power-amplifiers offer a number of potential advantages over their valve counterparts, particularly as regards size, weight and the amount of heat dissipated. It seems natural that the transistor should form a good power device for driving a loudspeaker because a transistor is basically a relatively low-impedance current-operated semi-conducting device whereas a valve is fundamentally a high-impedance voltage-operated vacuum-tube amplifier. Transistor circuits lend themselves much more readily to the elimination of the loudspeaker-matching output transformer, which is one of the more bulky, expensive and critical items in valve amplifiers.

On the other hand, power transistors introduce a number of new problems of their own. They are inherently more non-linear than valves and are more likely to be damaged by temporary current or voltage overloads. As the active parts of transistors are very small in area, it is difficult to dissipate any internally generated heat rapidly enough to avoid overheating and damage. They have to be intimately connected to metalwork, preferably with good heat dissipation in the form of fins etc, ie heat sinks. The heat dissipation problems caused by the relatively high standing current required by class A circuits and their comparative inefficiency has caused push-pull class B output circuits to be favoured for transistor amplifiers. Two particular problems arise here. Firstly, the amplifying transfer characteristic of transistors tends to follow a more nearly exponential curve than the equivalent valves. This means that the initial part of the characteristic near the zero point shows more origin curvature or "bottom bend." This tends to cause much greater "cross-over" distortion in the centre area of the combined transfer characteristic which is obtained in class B transistor push-pull output stages. Also, as transistors require an appreciable input-drive current, the position is again worse than for valves, which require virtually no input-drive current. The bottom bend curvature of transistors causes the effective input resistance to vary during the drive cycle. This means that the previous transistor drive stage must represent a low-source impedance in order to avoid distortion due to the drive current varying during the cycle in a non-linear fashion. One solution is to use a transformer-coupled drive stage, but this is contrary to the general desire to eliminate transformers from transistor circuits as far as possible, on the grounds of cost, bulk and the small amount of non-linear distortion which is inherent in all iron magnetic circuits. It is also difficult to avoid some lack of balance between the two halves of the push-pull transformer windings, unless bifilar or multi-wire windings are adopted. The application of a large amount of negative feedback over the output and driver stages will reduce the distortion to a low figure, but there is still liable to be some detectable audible distortion of low-level signals which are in the high distortion region around the cross-over point between the two output transistors, unless special precautions are taken.

Unfortunately, it appears that this type of distortion on low-level signals, such as quiet voices or quiet unaccompanied string passages, represents a relatively large amount of the higher-order harmonics, typically third harmonics and higher. These are objectionable to a much greater degree than their measured amounts might suggest. Factors of this sort have tended to cause transistor amplifiers to sound inferior to equivalent valve amplifiers, even though measurements of total harmonics might appear to be similar. Also distortion measurements are often quoted at full output, not for low levels, though some manufacturers do quote total harmonics at 10 milliwatts as well as at full rated output. Genuine comparisons should ideally quote these as well as intermediate levels.

It is possible to reduce all distortion to a suitably low level and to give results equal to the best valve amplifiers if careful circuit design and component matching is applied to transistor drive and output stages. Full complementary working of matched silicon output transistors and the achievement of constant-current drive and correct balance over the whole frequency range are among the methods adopted by the leading high-quality manufacturers. Good regulation of the power supply is also necessary to avoid voltage drop on sustained loud signals.

Precautions against output transistor breakdown and damage

Valve amplifiers are largely immune to accidental electrical damage if they are correctly designed and the components are properly rated. Thus overload by high inputs, surges, short-circuits or accidental disconnection of the loudspeaker under load etc cause no damage.

Transistors are much more vulnerable to permanent damage due to current or voltage overloads, and built-in precautions should be incorporated, either in the form of overload trips, fuses or automatic limiting circuitry. The possible combinations of overload conditions are not always easy to predict and considerable development has been carried out by the leading manufacturers. Their amplifiers are probably proof against most forms of accidental overload, due to the use of such devices as sensing transistors to monitor output transistor emitter currents or output voltage levels, the sensing circuits being made to operate quick-acting diode switches. This is a more sophisticated approach than the earlier latching overload trip relays. These generally protect against current overload only and may have to be specially designed in order to ensure fast enough operation. Unlike advanced electronic protection circuits, they usually put the amplifier out of action until they are re-set manually, the power supply being cut off.

If simple unprotected power transistor circuits are used, it is desirable to observe certain precautions. Before use, input and output terminals, plugs, wiring etc should be checked to ensure that no intermittent or short-circuited connections are likely. It is not always advisable to connect unusual output circuits involving excessive reactance, such as long extension leads, equalizers, piezoelectric or electrostatic units, in place of the normal moving-coil loudspeaker speech-coil that the circuit has probably been

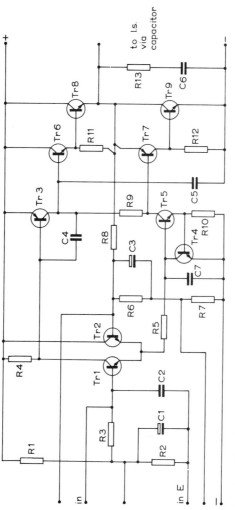

Fig 34. The schematic of a modern high-quality transistor power-amplifier module for Hi-Fi sound reproduction. Two complementary silicon driver transistors are used in a highly developed circuit using origin offset and constant-current drive with a large amount of negative feedback, to ensure that the non-linear distortion at both high- and low-output levels is kept at very low figure; there are no transformers and the unit is accommodated on a compact printed circuit board.

designed to operate. Other reactances may upset feedback circuits operating over the output stage and cause ringing or even inaudible high-frequency oscillation. As many transistor circuits will operate efficiently up to radio frequencies, there is a risk of unsuspected thermal damage from this cause. If a separate external output transformer is used for any purpose, such as extension loudspeakers, earphones etc, and the circuit should happen to be broken whilst carrying a high signal level, the back-emf generated in the high inductance of the transformer winding may be large enough to cause voltage breakdown of the output transistors.

It is, of course, inadvisable to cause excessive surges by changing input plugs etc without first turning the volume control to minimum, as these may also damage loudspeaker diaphragms and suspension if the low-frequency response of the amplifier is unusually extended, as well as possibly initiating thermal overload conditions in the output transistors.

Practical amplifier considerations

We see that the big advantages of transistors lie in the enormous reduction in size and heat dissipation and in potentially increased life and reliability.

The transistor is basically a low-impedance bipolar current-amplifying device as opposed to the valve, which is basically a high-impedance unipolar voltage-amplifying electron tube. The term "bipolar" means that the current is conducted in a normal junction-type transistor both by means of electrons and by "holes" or electron deficiencies within the crystal. A "unipolar" device conducts by means of electron movement only. All electronic valves and certain specialized semi-conductors (eg field-effect transistors) are of the latter type. We have noted that the inherently low impedance of semi-conductors as compared to the high impedance of vacuum tubes means that both input and output transformers can be largely dispensed with. This is a great advantage as regards size and cost and the small inherent non-linear distortion produced by any magnetic "iron" cored component is eliminated. This particularly applies to output transformers, where the distortion due to magnetic core saturation may be by no means negligible, unless it is offset by the application of a fairly large amount of negative feedback over a path including the output transformer and also over some or all of the power-amplifier stages. The distortion produced by input transformers is very much less, due to the minute current swing produced by a microphone or low-impedance pick-up.

Transistors have two major drawbacks from the point of view of the circuit designer: their operating characteristics are more non-linear than those of valves, and the "production spread" or variation in characteristics from one transistor to another is also often much greater. This means that for very low-noise input-stage applications or where the maximum efficiency with low distortion is required in output stages, transistors may have to be specially manufactured or specially selected, with a consequent increase in cost.

Thus, in spite of the possible simplifications and the reduction in size, the cost of really high-performance transistor amplifiers may not be much less than that of valve amplifiers.

Performance specifications and functions of amplifiers

The basic function of an amplifier is, of course, to increase the minute amount of electrical power available from microphones, pick-ups etc to a value which will give satisfactory sound reproduction from a loudspeaker system. Subsidiary functions include switching, response equalization (tone control), volume control, impedance transformation (eg matching of a high-impedance device such as a crystal 'pick-up) provision for the elimination of unwanted non-essential signals from the reproduction (eg turntable rumble, surface hiss on old records, radio interference etc).

The specification and design of an individual amplifier to perform one specific job can be simplified and restricted to the barest essentials required for efficient operation in the required manner. This might result in the cheapest possible design, but it might be found inadequate to perform other tasks for which the need might arise at some future date.

Thus it is usually found that, except in the very cheapest amateur-built designs and the cheaper commercial products for home sound reproduction, provision is made for some flexibility in the design with regard to such important parameters as the number and type of inputs accepted, possible variations of gain and response etc.

The power output of an amplifier must be decided, having regard to the efficiency of the loudspeakers, the size of the room and the type of programme to be reproduced.[3] To take two extreme examples, a bass-guitar amplifier would require a large power reserve in order to accept the enormous low-frequency transients fed into it and is normally required to be reproduced in a moderate-sized hall, whilst the reproduction of eighteenth-century chamber music in a small living room with an efficient high-flux loudspeaker would need only a comparatively small power output.

It is obviously sensible to allow an adequate margin in the power output of a given amplifier and, in fact, the reproduction of most types of speech and music demands a peak-power handling capacity very greatly in excess of that needed to handle all but the peak programme levels. For most domestic purposes a very general rule of thumb might be that 3 watts available power is about the minimum which should be considered for high-quality reproduction, whilst between 10 watts and 20 watts should be sufficient for all but the largest rooms.

The common reference to "available power" brings us to one of the first difficulties which the amateur encounters when trying to assess the relative merits of the various types of amplifiers which he can choose from. It is, in fact, difficult to specify the actual useful power which an amplifier will feed to a given loudspeaker under practical conditions. This power figure depends on a number of variable factors. One of the most important is the actual motional impedance value of the loudspeaker and its relative resistive and reactive values at any given frequency. This may be very different from the nominal impedance value quoted for the unit concerned. Difficulties of this sort have led to amplifiers often being rated in terms of the "steady-state" power dissipated in a resistive load equal in value to the nominal loudspeaker impedance which the amplifier is designed to feed. This is what is usually implied when an amplifier power is stated. The usual rating calls for amplifier power to be specified in this way. Other amplifier ratings are given which specify a peak programme or music power rating which gives higher figures for amplifier powers (peak power is $2 \times$ r.m.s., IHFM music power is $1 \cdot 5 \times$).

The above observations should be borne in mind when comparing amplifier specifications, together with the fact that a good power rating at middle frequencies does not necessarily mean that the performance is satisfactory at extreme low and high frequencies. The situation is further complicated by the fact that it is not possible to run some transistor amplifiers continuously at their full programme rated output, on account of overheating of the output transistors. Normally, there is no real risk of burning out the power transistors on programme because of the relatively enormous peak-to-mean-power ratio encountered on all usual programmes.

In fact, the design and full checking of modern efficient high-quality amplifiers demands a very considerable amount of expertise. Complete checking involves measurements of noise, pulse response and non-linear (intermodulation) distortion. The amateur obviously cannot carry out tests of this kind without having a lot of expensive professional test equipment at his disposal. He can, however, buy a good reputable amplifier or kit or build to a good published design with considerable confidence in the available power provided.

Amplifier controls are very important and their operation must be properly planned so as to be logical and convenient. This is an aspect where ideas vary and there is considerable latitude for experiment.[15]

Typical high-quality amplifier specifications
An indication of good practice can be obtained by examining the design of representative power-amplifiers, and preamplifiers. A monophonic or single-channel chain is usually first considered. Stereo would demand duplication of the channels with probably a larger common power-supply unit, though the power output requirements per channel could normally be approximately halved for domestic use.

A figure of about 12 watts for a monophonic power-amplifier is generally found to be adequate for a good domestic installation. This power is considered to be the resistive load value, as measured.

The detailed specification of a preamplifier/control unit or a power-amplifier in terms of performance figures gives a fair guide to the merit and capabilities of the equipment, but certain reservations should be borne in mind. Firstly, the exact meaning and validity of the figures depends on the method of measurement and the instruments used. The example of power-output measurements mentioned illustrates the point. The dilemma applies to other pieces of audio equipment and stems from the fact that speech and music represent extremely complex signals and thus instrumental (electrical) measurements are necessarily simplified and to some extent artificial. They have been framed to give as

Fig 35. *A picture of the power-amplifier module of Fig* 34. *The printed circuit board is* $2\frac{3}{16}$ *by* $2\frac{1}{2}$ *in. The power output transistors at the top of the picture are mounted on a matt black heat sink large enough to dissipate the heat generated by moderate music powers. Larger powers may be dissipated if the heat sink is mounted intimately in contact with a larger metal chassis etc.*

much information as possible, but cannot always supply a complete comprehensive picture of all aspects of amplifier performance. Any amplifier tests should therefore be backed up by special transient or square-wave tests and by listening tests on specially chosen programme material.

With these reservations in mind we might specify a typical high-quality amplifier channel as follows—

FREQUENCY RESPONSE AND STABILITY CRITERIA
Radio, tape and microphone input response should be ± 1 dB from 50 Hz to 15 kHz, ± 2 dB from 20 Hz to 50 Hz and from 15 kHz to 20 kHz, for flat response position of controls. Any response outside the extremes of 20 Hz and 20 kHz is not

normally an advantage and may actually cause trouble by re-producing sub-audible turntable or microphone rumble etc below 20 Hz and radio carrier or bias oscillator leaks etc above 20 kHz. Although not directly audible, interfering signals of this sort can cause considerable trouble by overloading various parts of the system and possibly causing severe whistles or intermodulation distortion, in addition to introducing a risk of overheating transistor output stages etc. However, the desirable frequency response of an amplifier in the top cut-off region is intimately associated with its phase response, that is the extent to which the angle of the signal vector lags behind or leads the signal at lower frequencies. If an attempt is made to give an amplifier a very sharp response cut-off immediately outside the audible range this is bound to cause a violent change in the phase response in the cut-off region. The effect is to cause a damped oscillation to persist at the frequency in question. This will be excited by any electrical transient and may lead to "ringing" or a semi-continuous oscillation. A slightly more severe phase shift may in some cases actually produce a positive feedback path around some path of the amplifier. If some gain due to a tube or transistor etc occurs in this path, continuous oscillations will build up and the amplifier is then "unstable." A very complete circuit theory has been worked out for amplifiers involving negative feedback. This theory has a fundamental bearing on high-quality amplifier design, in that the application of a fair amount of negative feed-back is essential in order to reduce non-linear distortion to a low level as well as to control the frequency response. The theory shows that if a polar plot of the gain and phase characteristic around a circuit loop is carried out, the system will be stable unless the polar curve encloses the point $(+1, 0)$. This is the well-known "Nyquist criterion," which determines the stability of all feedback systems, including all types of electronic and other amplifiers, and also electro-mechanical servo-systems such as high-speed voltage recording test instruments (eg Brüel and Kjaer level recorder) or large controlled dynamic structural vibration-exciting equipment. If the transmission properties of the path vary with signal level so as to cause the Nyquist plot to enclose the critical point on overload etc but not for normal signal levels, the system is only "conditionally stable." A good amplifier is stable under all conditions of load or temporary overload and is said to be "unconditionally stable." The rate of cut-off of the response outside the pass-band is bound up with the phase response and, in general, should not have too rapid a fall. If steep-cut filters, cross-over networks etc are desired, these generally should be introduced between amplifier or buffer stages so that they are not embodied in any part of a feedback loop. The whole system must still be carefully investigated with the aid of square waves or other transients in order to see whether any appreciable amount of "ringing" has been introduced. The application of large square-wave pulses at various repetition rates or frequencies and examination of the output waveform will show if the amplifier is running into oscillation on overloading peaks, ie is only "conditionally stable." A good check of stability for an amplifier with an output transformer within the feedback path (ie negative feedback taken from the loudspeaker or output terminals) is to apply a 30 Hz square-wave input signal large enough to produce a degree of overloading. The output waveform should not exhibit bursts of higher frequency oscillation at any part of the low-frequency cycle.

Enough has been said in the preceding paragraphs to show that the design of a high-quality amplifier to have a particular frequency response is not a simple matter and that a number of quite sophisticated tests should really be applied to see that the amplifier is stable under all conditions, in addition to having the desired frequency response.

When the negative-feedback loop of an amplifier is being designed or adjusted, various "tricks of the trade" can be used in order to avoid the necessity of analysing the circuit in detail on paper with full knowledge of all relevant stray or leakage circuit elements, or of using expensive professional equipment such as a Nyquist diagram plotter. One artifice is to open the feedback loop at a convenient point and measure the overall frequency response around the loop. The rate of fall of the response above the cut-off point should not usually greatly exceed 12 dB/octave for a good stability margin to be achieved.

GRAMOPHONE INPUTS ETC

These should be equalized to within $\pm 1 \cdot 5$ dB of the BS or RIAA standard record equalization curves for $33\frac{1}{3}$, 45 and 78 rpm on the assumption that pick-ups with ideal frequency response are used. The appropriate input selector switch positions should be handy and clearly visible.

BASS AND TREBLE CONTROLS

These controls should be arranged to give about ±15 dB maximum lift or loss at both extreme bass and top frequencies. The response between the "starting points" of the boost or loss areas on the response curve should be flat and ideally the starting points should progress along the flat part of the curve according to the severity of the boost or cut provided. The well-known circuits due

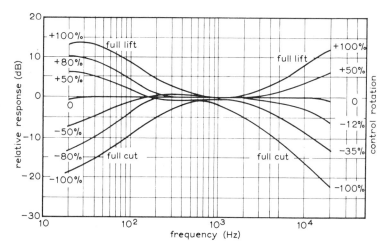

Fig 36. The range of independent continuously adjustable bass and top lift and cut tone control available with the well-known Baxandall active tone control circuit which may be used between two stages of a preamplifier.

to Baxandall have this desirable characteristic, which avoids undesirable exaggeration or deficiency of middle, low or high frequencies when control of extreme bass or top is sought. The bass and top positions should either be switched in an adequate number of discrete steps or be controlled by accurate and easily set continuously variable controls. The central flat position should be well defined and easy to find. Fig 36 shows bass and treble control responses of this kind.

RUMBLE FILTERS

It is not desirable to reproduce sub-audible frequencies due to the presence of turntable rumble, wind or breath surges on microphones etc. In order to get a steep and effective low-frequency

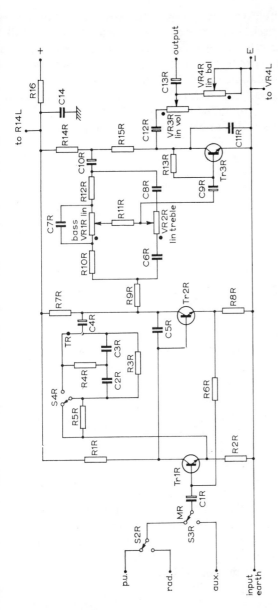

Fig 37. The schematic of a modern stereo preamplifier (one channel shown). The RC network between the first two transistors corrects the output of magnetic velocity-reponsive pick-ups for the standard recording characteristics of gramophone records. Continuously adjustable bass and top-tone control is provided by the potentiometers in the other RC network, which is sited between the second and third transistors. A stereo balance control is provided after the output volume control. All variable controls in stereo preamplifiers have to be ganged and matched to track accurately together.

cut-off at sub-audible frequencies without affecting audible low frequencies, it is necessary to use a multi-element scientifically designed and correctly terminated filter. This is liable to use relatively large and expensive inductances and capacitors. The former may also have to be carefully shielded in high-permeability screens of "Mu-metal" or Permalloy in order to avoid hum pick-up from the electromagnetic stray fields radiated by mains transformers etc. As it does not incorporate amplifying elements

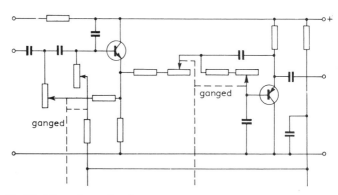

Fig 38. *The schematic of a stereo two-stage active filter unit (one channel only is shown). This may be connected between the stereo pre-amplifier and the power-amplifiers, so as to give independent bass and top steep-cut filters to enable low-frequency rumble and surges to be eliminated. A top steep-cut filter will reduce residual scratch, hiss and higher-order distortion products. The two sets of controls for the channels are accurately ganged.*

such as valves or transistors, this is a "passive" type of filter. A high-class filter of this kind can be permanently connected in circuit as the full audible bass response is retained. For more compact and less expensive amplifiers an altogether simpler form of rumble filter is preferable. An RC network incorporated in a negative-feedback loop around one or more amplifying stages can be made to give attenuation of 12 dB per octave below about 70 Hz, without much "roll-off" at higher bass frequencies. So-called "active" filters of this kind are usually arranged to be switchable "in" or "out" and are often referred to as high-pass filters. Fig 38 gives typical circuits for the two types of filter used, ie rumble and top-cut.

TAPE PLAYBACK EQUALIZERS

Standard tape-recording characteristics have been laid down by the CCIR (the European radio standardizing body). These characteristics differ for the various speeds 15, $7\frac{1}{2}$, $3\frac{3}{4}$, $1\frac{7}{8}$ inches/sec. If suitable equalizers are incorporated in an amplifier in addition to gramophone equalizers, the amplifier can be used directly with a tape reproducing-head for tape playback or for tape-recording monitoring from a separate head as a check whilst tape-recording is actually in progress (Fig 25). The $7\frac{1}{2}$ and $3\frac{3}{4}$ inches/sec tape-head equalization characteristics do not differ enormously from gramophone equalization characteristics and in emergency these may be used following a tape replay-head if some sacrifice in response is accepted. The proper equalization is normally an integral part of tape-recorder playback circuits, which produce a flat output.

TOP-CUT FILTERS

These are designed to give fairly steep high-frequency cut-off characteristics in order to deal with programmes containing an intolerable amount of high-frequency noise, hiss or distortion. They are also referred to as low-pass filters or steep-cut filters. As in the case of rumble filters, conventional multi-element passive filters may be introduced between amplifier stages to give high rates of cut-off. As we have seen, if the rate of cut-off is too

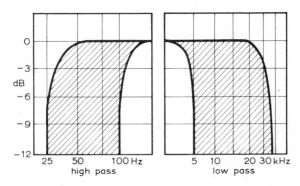

Fig 39. *The range of responses available from the active filter unit shown in Fig* 38. *The shaded areas show the range in which the steep-cut filter characteristics are independently adjustable. In the widest response position of the controls, the system is flat between* 40 *Hz and* 20 *kHz, the narrowest response being between approximately* 120 *Hz and* 4·5 *kHz.*

high, "ringing" or severe transient distortion near the cut-off frequency may be introduced and so in practice the rate of cut-off is likely to be restricted to about 18 dB/octave maximum. RC networks used as feedback elements in an active filter circuit are now used to provide top-cut filters and will readily produce cut-off rates of 12 dB/octave or more, which is often sufficient for any but the most severe hiss, even from elderly shellac records etc. Popular values of nominal cut-off frequencies for switched filters are 4 kHz, 7 kHz, 10 kHz and 12 kHz. In some cases, the cut-off frequency or turnover is infinitely variable over a similar range by using a potentiometer as one of the resistance elements in an active filter circuit.

Equalization characteristics for older records

The majority of gramophone records made since 1954 are recorded to the standard pre-emphasis curve described by the three time-constants 3180, 318 and 75 microseconds for $33\frac{1}{3}$ and 45 rpm records. 3180, 450 and 50 microseconds were used for some 78 rpm records. For records issued before 1954, various recording characteristics were used and some trial and error may have to be used for best reproduction. A number of earlier records are recorded to various specified curves and some makers have published correct equalization characteristics for makes and labels issued prior to 1954, but they warn that shared or otherwise unspecified masters may have been used in some cases.

Parallel operation and unconditional stability of amplifiers

We have noted some of the factors affecting the stability and frequency response of amplifiers and the manner in which negative feedback has a bearing on the amplifier characteristics. Virtually all valve power-amplifiers, as distinct from transistor power-amplifiers, have to use an iron-cored output transformer in order to match the relatively high-impedance valves to low-impedance speech-coils. If amplifiers are to have closely controlled response, gain and distortion characteristics, a fairly large amount of negative feedback has to be applied over several amplifier stages, including the output transformer. This places very stringent requirements on the design and construction of the output transformer with regard to the phase shift introduced by circuit elements such as leakage inductance and winding stray capacity. It is possible to minimize these factors by subdivision of the various windings and by ensuring exact symmetry in winding and connecting the various coils. A good power-amplifier is stable with all types of load and a number of amplifiers may sometimes be operated in parallel without risk of instability if a larger power output is required. Generally, however, it is not advisable to attempt to run amplifiers in parallel in order to get more power output unless they have been designed and tested with this in mind.

Stereo amplifiers system matching

Ideally the two stereo chains should be accurately matched as regards gain, phase, frequency characteristics etc. Overall gain should generally be matched to better than 2 dB. Commercial stereo amplifiers are provided with accurately matched and ganged controls for switching, bass, treble, volume etc. In addition a balance control is provided giving a fine control of the relative gain between the two channels. Some amplifiers have been produced with dual concentric knob controls, so that the characteristics of the two channels can be individually varied at will. A phase-reversing switch to give reversal of polarity on one channel used to be a necessary feature in the early days of stereo records when record channel phasing was not standardized. This facility is still sometimes provided on amplifiers, although once correct phasing has been established it should not nowadays be necessary to use a reversing switch.

Stereo listening on earphones

The use of matched high-quality earphones for binaural listening to stereo programmes can be very well worth while, and earphone outlets in the form of special jack sockets or DIN outlets etc are often provided on amplifiers. Moving-coil earphones are often wound to about 300 ohms and may be shunted across loudspeaker outlets of 3 to 15 ohms, dummy-load terminating resistances preferably being provided if the loudspeaker is disconnected. Ganged earphone volume controls and "cross-mixing" facilities are additional refinements which may be added for earphone listening. Cross-mixing circuits are sometimes provided to dilute the left-right stereo separation, which some people may find to be excessive on stereo earphone listening (Fig 61).

Note that safety precautions may be necessary when connecting earphones to equipment. Again, a dealer should be consulted.

7

Loudspeakers and enclosures

The loudspeaker is one of the most important items in the reproducing chain. In addition to its electro-acoustic role, it must be considered aesthetically as something which is necessarily an important item to be fitted in your furnishing scheme. Where stereo is concerned, the problem may be even more difficult to resolve. Thus the loudspeaker is one of the most difficult units to choose when deciding on a reproducing system.

It is, of course, possible to go for a monolithic system in the form of a single cabinet which houses the loudspeaker as well as the amplifier, turntable, tuner etc. Most enthusiasts, however, will prefer to have their loudspeakers housed in individual cabinets, as this allows the maximum flexibility in both performance and placement of the loudspeakers.

Very few will wish to attempt to construct their own loudspeaker motor units, though this is not impossible. Many will certainly wish to make their own enclosures or cabinets, as this can be both satisfactory and economical to undertake. It is then first necessary to decide on the type of loudspeaker units to use. These range in price literally from around £1 to scores of pounds, and an optimum choice for any given projected system can prove very difficult to make.

Not only are loudspeakers difficult to apply correctly, but they are probably more prone than any other part of the system to introduce subtle forms of distortion and coloration of the sound. These are not only generally undesirable in themselves but may be very difficult to predict from response curves and casual listening tests. The art and practice of loudspeaker measurement is not generally as advanced as that of amplifiers and other parts of the system, and it is extremely difficult to make reliable comparisons between loudspeaker units, particularly at different times and different places.[16,17]

136

It is pertinent, therefore, to list a few guiding principles when first considering the choice of a loudspeaker unit.

Choosing a loudspeaker unit

In the first chapter we discussed the general way in which loudspeakers perform in rooms and obtained some idea of what to expect from the various types. Later in the present chapter we will go more deeply into the design and construction of various types of loudspeaker and cabinet. It is, however, possible to assess and to select the type of loudspeaker best suited to one's own particular needs by considering a number of general rules. Once the field has been narrowed, it then becomes easier to decide between the remaining alternatives. Even when the more technical details of performance are available in the form of response curves, distortion characteristics, transient response criteria etc it is still extremely difficult to arrive at a comprehensive "figure of merit" for a loudspeaker. The final choice for most people is thus to some extent a personal decision based possibly in part on cost, on such listening tests and assessments as one is able to make and on the recommendations of reviewers, experts and friends.

Firstly, one should consider the exact purpose for which the loudspeaker is required, for example—

1 Low, medium or large power output. This is only partly dependent on the size of room. The dominant type of programme and use should be considered. For example, electrical instruments such as electric guitars, electronic organs etc are liable to give rise to sustained low-frequency notes of such power that the loudspeaker units must withstand larger cone excursions and speech-coil heat dissipation than normal.
2 Whether the unit is to cover the whole frequency range or only a part of it. The latter category will include bass woofers, mid-range units, high-frequency or ultra-high-frequency tweeters.
3 The type of cabinet or enclosure in which the unit is to be housed.
4 The impedance of the unit and the way in which it is to be mounted and electrically coupled to any other loudspeaker units in the same array.

Additional decisions involve the operating principle and whether the unit is a direct radiating system or horn-coupled. The great majority of loudspeakers today are direct radiating moving-coil types, with the exception of wide-range electrostatic units such as the Quad. Ribbon, electrostatic and ionic tweeters are also available. Electrostatic and most moving-coil tweeters are direct radiating types, but the other two types mentioned above are horn-coupled.

A final decision is the basic efficiency desired. In some units this is fixed by the designer but many of the more widely used direct radiating moving-coil units are available in low, medium and high magnetic flux densities. In these days of highly efficient transistor amplifiers, it is not so important to strive for the utmost in efficiency as it used to be, and medium-flux units are often quite satisfactory (about 9000–11,000 gauss). Units with the maximum flux density will generally have slightly better damping and a better transient response, with the result that they are often specified for the highest quality.

The cost of moving-coil units depends on the size, the complexity or any special treatment of the diaphragm system, the flux density and, a most important factor, the quantity in production at any given time. Hence lower-flux units for the mass-produced radio and TV market may be obtained quite cheaply. These may sometimes be used successfully in high-quality arrays, particularly when they are only required to cover part of the frequency range. However, it must be borne in mind that the response characteristics of these units are often tailored to suit a fairly restricted frequency range. A fairly high bass resonance is usually provided, partly as matter of economics, in that it is easier to provide robust cone surrounds on a mass-production basis than to make the necessarily more critical flexible edge rolls or other forms of termination required for the highly compliant cone systems, which must be used in efficient "long-throw" loudspeaker units, combining the properties of reasonable size and a low bass resonance. It is also undesirable to reproduce extreme low frequencies in the cheaper radio and TV sets in order to avoid placing too stringent a requirement on the level of mains hum, induced frame pulses, acoustic feedback to turntables and pick-up cartridges etc. Sideband-cutting may also sometimes be compensated by a broad resonance between 3 and 6 kHz in AM receivers, whilst inter-station carrier whistles may be reduced by means of a dip in the loudspeaker response at around 9 kHz.

In high-quality systems, broadcasts are preferably received on

VHF FM, and optimum electrical circuitry and loudspeaker cabinet arrangements are used; it is obviously desirable to use specialized loudspeaker units designed to produce a flat response over the widest possible band. The cheaper types of mass-produced units obviously have limitations in these respects, unless they are very carefully applied so as to utilize the flatter parts of their frequency range.

Generally speaking, the same axiom applies to loudspeakers as to other transducing elements, such as microphones or gramophone cartridges. This is that one should, in general, spend the maximum one can afford on these items because the range in performance between the cheapest and the most expensive is almost certainly much greater than that applying nowadays to other items such as amplifiers, cabinets etc. Furthermore, it is possible to save much money by constructing these latter items oneself, which is emphatically not the case with transducers unless you are exceptionally well qualified and equipped.

When deciding on the purchase of loudspeaker units the problem that is most difficult to resolve, apart from power handling, is probably the question of exactly how far one wishes to go in the reproduction of extreme low or extreme high frequencies. Assuming that you have considered the power-handling capacity of your amplifier at the frequency extremes, it is then necessary to examine the implications both in size and cost of providing a given extension of frequency range at either extremity.

It may, in fact, be debated how far it is necessary or desirable to reproduce the lowest and the highest octaves at the limits of audibility. For present purposes we may define the lowest octave as 30–60 Hz and the highest as 10–20 kHz. By reproduction, we imply reproduction at full strength. It must be borne in mind that really full low-frequency reproduction of this kind can add considerably to the size and cost of the loudspeaker units and the cabinets, to say nothing of the more stringent power output requirements of the amplifier. Full high-frequency rendering of this kind probably means not only using special tweeters but may also involve special measures to ensure adequate sound distribution or polar response.

There is no doubt a case for providing such a response if perfect reproduction of sound is to be attempted but, in practice, on a great many programmes and systems, the amount of sound energy present at these extremes may be small and, moreover, there may be undesirable extraneous or intruding noises which are so disturbing as to make it necessary to restrict the frequency range of the final reproduction.

Limiting factors in wide-range loudspeaker reproduction

If the bass response is extended to be really flat to the lowest audible frequencies, it is true that there may be enhanced enjoyment of certain musical sounds if, in fact, these are present in the incoming electrical signal. Actually, sounds of extremely low pitch are liable to be obscured by unwanted acoustic surges, due to the effects of draughts and air currents on microphones, residual traffic rumbles etc, and hence this part of the spectrum is liable to be removed prior to transmission or recording, so as to avoid over-modulation and distortion. In our own equipment, when reproducing records, turntable rumble and vibrational excitation of the very low frequency resonance between the tone-arm mass and the small restoring stiffness of a high-compliance cartridge may also generate an output of extreme low frequency. If this is reproduced at an audible level, we would normally resort to a bass-cut rumble filter, typically giving an attenuation of 12 dB per octave below various nominal cut-off frequencies switched to occur at 70 Hz and below.

Thus, for various reasons, most people do not in fact listen over long periods to extended low frequencies. The question one has to consider is whether it is worth the considerable cost and elaboration of having such a loudspeaker system. The answer is generally some form of compromise, with the smaller types of bookshelf loudspeaker at one end and the necessarily larger and more expensive loudspeakers at the other end of the scale. The efficient reproduction of extreme bass is costly in money and in space. The user has to decide whether it is worth it, and, if not, just how much bass he does wish to have available from his loudspeaker system.

At the high-frequency end of the spectrum similar arguments may apply, in that it is often necessary and desirable to cut part or all of the upper two octaves, so as to minimize the often very considerable annoyance caused by residual higher-order distortion products and high-frequency noise and hiss from a wide variety of causes, ranging from radio static to thermal and other noise in electrical circuits, to say nothing of magnetic-tape noise and record hiss.

However, given a really clean programme signal, there is no doubt that an extended high-frequency response to a point well above the normal pure-tone audible limit is essential to give completely realistic reproduction to transients and other sounds rich in high-frequency components.

Really good programme conditions of this sort are increasingly available today in FM reproduction of live broadcasts in areas of good signal strength, and in both tape and disc recordings using modern methods of noise reduction and means of controlling the recorded level.

The provision of ultra-high-frequency tweeters is not particularly expensive in terms of cash or space, and so it seems reasonable to consider the provision of a flat and extended high-frequency response, as long as this is balanced by a reasonable low-frequency response.

Polar response considerations

One further parameter to be considered when choosing loudspeaker systems is the type of directional response which can be obtained from different designs. Enclosed types of cabinet loudspeakers are basically omnidirectional at low frequencies but the units become directional at high frequencies where the wavelength of sound is small compared to their size. Tweeters thus require to be small in order to give a well-spread beam of high-frequency sound.

Open-backed loudspeakers, such as wide-range electrostatic units, are fundamentally bidirectional with zero output in the plane of the diaphragm. In fact, it has been claimed that a good stereo effect is obtained if the figure-of-eight polar response is maintained up to high frequencies, together with a less critical listening position.

On the other hand, it may be found that the positioning of bidirectional loudspeakers is more critical in some rooms, and this should be borne in mind as a possibility when considering their choice. If in doubt, it may be possible to arrange a demonstration in your own listening room. This should not be hurried, as the excellence of the results which may be obtained justifies taking trouble to find the best siting and also listening to a wide range of programmes over a period of time.

In fact, this applies to listening tests on any new loudspeaker system. Established listening habits die hard and one may have to live with a new system for a while before appreciating the results completely.

The moving-coil loudspeaker unit

We have considered some of the main factors involved in the choice and use of the various types of loudspeaker likely to be used in home sound reproduction. We will now consider the design and the basic mechanism of loudspeaker units in more detail. The moving-coil loudspeaker driving unit is by far the most popular type in use today, but we have noted that other principles are used. The push-pull double-sided electrostatic system is used successfully for high-quality wide-range loud-speakers, and to some extent as tweeters. Ribbon loudspeakers are also used as tweeters, as also are ionic loudspeakers. It may be noted that the driving force in the case of moving-coil units is on the speech coil only and thus is at the apex of the cone. In the other types mentioned the driving force operates over the whole active radiating area, in other words, over the whole of the current-carrying ribbon or over the whole of the charged diaphragm surface of the electrostatic unit. In the ionic unit the force is generated due to thermal expansion of the electrically charged air particles in the sound-originating cell.

The general principles and constructional layout of the almost universal moving-coil unit have not altered very much since it was first introduced some forty-five years ago. The driving force is due to the direct interaction between the signal current and a constant radial magnetic flux in the gap and it is basically free from non-linear distortion. A conical or flared shape is given to the diaphragm in order to make it as light and as rigid as possible, so that the efficiency of sound radiation may be kept as high as possible. A flexible edge surround as well as a radial centring "spider" or a bellows member are usually required to keep the coil centred in the gap.

The exact construction of the various types of moving-coil loudspeaker unit naturally depends to some extent on the size, the frequency range covered and the type of magnet used.

Several typical units are illustrated in this section of the book. In the case of the electrostatic loudspeaker, the diaphragm does not have to be rigid and thus can be both light in weight and of a large area. In contrast, the cone of a moving-coil loudspeaker has to be relatively small in order to meet the need for rigidity, about 18 in. being the maximum size for bass units; wide range units seldom being more than about 12 in. effective diameter.

Modern units use cast or sintered permanent magnets producing a flux density of between 7000 and 17,000 gauss in the gap. A die-cast or pressed-metal chassis is attached to the magnet and is extended forward to support the diaphragm surround and to form the mounting ring for attaching the loudspeaker unit to the cabinet front. The diaphragm is usually made from a special felted paper composition or, in some cases, of a light metal, a reinforced rigid foamed plastic or a solid plastic material. Speech coils are wound from copper or aluminium wire on varnished paper or plastic formers.

The moving system at its simplest may be considered as a rigid mass suspended on a restoring spring. An alternating driving current will give substantially the same driving force at all frequencies, but resonance will occur between the mass and the spring at a frequency normally arranged to be at the bottom end of the frequency range covered by the loudspeaker. Above this point the mass effect increasingly predominates over the spring as the frequency rises. The sound radiated from a rigid cone can be calculated by taking it to be equivalent to a flat circular piston mounted in a large flat baffle. The result shows that the acoustic radiation load offered to such a piston increases with frequency up to a distinct cut-off point. The air moved by the surface can be considered as having both mass and resistance components. The mass component adds slightly to the diaphragm mass and represents kinetic energy stored in the air at points in the cycle where the velocity is at a maximum. Some of this kinetic energy is returned and stored as potential energy in the spring at points representing spring deflection. The resistive component represents the sound energy radiated and is the useful component producing the sound output. The sound energy radiated by such a mass-controlled piston should be substantially constant between the bass resonant frequency and the top cut-off frequency. The response on the axis is slightly improved at higher frequencies because the sound energy becomes concentrated along the axis at the expense of off-axis sound. It is evident, however, that a strictly rigid diaphragm would not be able to cover the full frequency range, but in practice very few cones are completely rigid and in fact most are given a degree of controlled flexibility which allows a number of subsidiary break-up modes of vibration to occur in addition to the main bass resonance. Ideally it is possible to extend the response well beyond the theoretical top cut-off point, but obviously at some frequencies parts of the diaphragm will move mutually in-phase whilst other parts may move in anti-phase. The result is likely to be a response showing a number of peaks and dips. The whole effect is very complex, but with good design a satisfactory wide-range response may still be obtained.

When deciding the size of cone to be used, we note that a large cone can move a given volume of air with a smaller excursion and with less non-linearity, but it may rely to a greater extent on higher modes in order to cover the frequency range, and it may become excessively directional at high frequencies. Ideally, the size should be tailored to the frequency range the unit is required to cover. This is, in fact, what we usually find in the case of high-quality multi-unit loudspeakers, where we have woofers, mid-range units and high-frequency tweeters in a descending order of size. One additional point, however, is that for a given output a larger cone is liable to require a larger size of cabinet with a consequent increase in cost. This has lead to an improvement in the performance of smaller mid-range and wide-range units, particularly as regards the permitted excursion of the cone. Edge surrounds and centring bellows behave as non-linear restoring springs at any excursions exceeding a small fraction of an inch, unless very special precautions are taken in their design. Also turns of wire on the speech coil which move beyond the limits of the pole pieces encounter the reduced fringing magnetic flux and again tend to introduce non-linearity into the force on the coil. This can be minimized by careful proportioning of the pole pieces and by such devices as making the coil shorter than the gap depth. An idea of the magnitude of the problem is obtained when it is realized that a small cone may have to move through a total distance of $\frac{1}{2}$ in. to reproduce a bass note at full concert hall strength in a large living room. The effect of non-linearity is, of course, not only to produce harmonics of single tones but also to produce much more objectionable combination tones when many frequencies are being reproduced simultaneously, as in normal programmes. Additionally, sub-harmonics as well as harmonics and combination tones may also be produced by flexing of the cone material. The overall result is that the actual value of non-linear distortion varies widely at different frequencies and can generally only be given conveniently in the form of a continuous curve. Another effect, known as Doppler distortion, occurs when a loudspeaker is simultaneously reproducing strong low- and higher-frequency tones. The high frequency will be increased or decreased in pitch according to whether the cone is moving back or forwards at the low frequency. This form of distortion is inevitable in wide-range loudspeakers but is fortunately not usually very severe. It is greatly reduced if the range is divided between a number of units.

The construction of moving-coil units

We have seen that in detail this is determined by the cone size and shape, by the magnet and, to a large extent, by the permitted cost of the unit. Cones may be circular or elliptical in plan with a "curvilinear" or a straight-sided section. Large or small integral-edge corrugations may be used or large single-edge rolls may be cemented on to the cone edge. Alternatively, foam or fabric edge materials may be used. Integral-edge corrugations may be treated with viscous compounds in order to add damping and to absorb the mechanical waves traversing the cone, and thus to prevent "standing waves" being set up along the cone section. The centring bellows are usually made as air-tight as possible so as to exclude dust from the outer coil gap. A fine-mesh dome made from stiff gauze is generally fitted over the coil end at the apex of the cone so as to seal the inner gap. This gauze allows some air passage so that the air inside the gap is not completely sealed off, as this might introduce an unwanted stiffness or might cause trouble due to atmospheric changes.

Magnets may be of the fully-columnar high-energy-content cast type, or may be of the large, flat ceramic ring type. The latter are cheaper but are larger in volume for a given gap flux density. Speech-coils are wound to the various standard low-impedance values (3, 8 or 15 ohms) and are connected via the rather critical flexible connecting leads. These must not fatigue or give rise to buzzes, and used at one time to be a considerable source of trouble. Some high-quality units make connection to the coil by means of a long flexible metal spring centring spider (Fig 40).

It is, of course, essential to ensure that the magnetic gap is permanently and accurately centralized. The older pole piece assemblies relied on screws, staking or welding, a special centralizing brass cup often being provided to ensure that the gap does not become eccentric if the unit is dropped or distorted. Such a

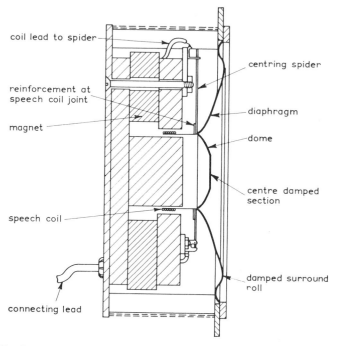

Fig 40. *A highly developed modern small wide-range loudspeaker unit. A high-flux ceramic magnet is used and considerable precautions are taken to avoid mechanical or magnetic non-linearity. A curved titanium cone is used with a highly compliant surround which permits large excursions in relatively small enclosures.*

magnetic system is in a state of unstable equilibrium, the tendency being for the force to pull the centre pole over in the direction of any small initial eccentricity. Epoxy resins are now largely used to secure poles and magnets together with a strong and permanent bond, ceramic magnets being particularly suitable for securing in this way because of their large surface area.

The electrostatic loudspeaker

It is known that the modern wide-range electrostatic loudspeaker may achieve a substantially linear driving force over a large flexible diaphragm mounted between rigid perforated electrodes closely spaced on each side of the diaphragm. The theory shows that the diaphragm must not only be polarized via a high resistance from a high-tension supply of up to several kilovolts, but it should also have an optimum value of surface conductivity. The ideal is to maintain a constant value of charge on the diaphragm during operation, without any tendency for the charge to migrate or move about on the surface during the AC cycle of operation. The precise manner in which the various requirements have been met in practice has been arrived at by many years of development by the manufacturers and undoubtedly represents a triumph of both research and the scientific application of materials. The achievement of mechanically non-resonant driving electrodes with correct acoustic transmission properties, as well as the permanent sealing of the whole system against the entry of dirt or moisture, is a considerable advance.

The loudspeaker is almost a pure capacity as far as the amplifier is concerned and it is necessary to ensure that the power-amplifier is designed to work into such a load without instability or ringing on transients.

Fig 41 shows the very successful high-quality wide-range loudspeaker made by Quad Ltd. This is a fairly large bidirectional

Fig 41. *The Quad electrostatic loudspeaker: a well-known wide-range unit of very high quality.*

unit which radiates a full bass response without the use of any subsidiary baffles or enclosure. Normally the higher frequency radiation of electrostatic units must be restricted by cross-over networks to a smaller narrow vertical strip so as to avoid undue focusing of high frequencies.

The extra complication, including the use of a relatively high-voltage polarizing supply, has so far restricted the use of electro-static loudspeakers to the higher-quality systems, though the possibility of permanently polarized "electret" materials becoming a practical proposition may remove some of the expense and complication.

The Ionophone

This is a very interesting invention as it represents a unique form of loudspeaker without any solid moving parts, as movement is directly induced into the air itself by means of a quartz cell at the throat of the horn, which is an essential part of this form of loudspeaker. The air in the cell is ionized or electrically charged by the application of a modulated RF high-voltage signal at a carrier frequency of 27 MHz. The effect is to cause variations in the heat of the air at the rate of the audio-frequency signal which modulates the carrier. The result is to send sound-pressure wave-fronts into the horn, which is matched to the outside air, so as to increase the sound volume in the usual way.

This is almost the only audio transducer in use which possesses no mechanical moving parts, with the result that the initial response is almost perfect, the properties of the horn representing the only limiting factor. The principle operates over the whole audio range and into the supersonic frequency range, the practical limitation at low frequencies being the size of the horn required subject to the limit of thermal linearity of the air in the cell. Messrs. Fane produce a very high-quality tweeter which operates from 3500 Hz to well over 20,000 Hz. It has an output which will accept the music power of a 20-watt amplifier in that range.

It must be noted that the electrical system of this type of loudspeaker is similar to the output stage of an AM radio trans-mitter, and thus it has to be carefully shielded in order to prevent unwanted external radio transmission. The Fane unit is claimed to be well shielded and, in addition, the 27 MHz band is chosen as representing an acceptable part of the radio spectrum for such purposes.

Moving-coil tweeters

We note that these may be horn-coupled or direct-radiating types. For domestic purposes where large amounts of power are not usually involved, the direct-radiator moving-coil tweeter has a number of advantages. The construction can be simpler and cheaper than that of a horn unit and the mounting in the main cabinet used for the other loudspeakers can be both simple and effective, provided that certain precautions are taken. The cabinet front is normally almost 1 in. thick. If a hole is bored in this and the tweeter mounted behind it in the same manner as for other loudspeaker units, a cavity resonance may occur in the high-frequency range and may produce audible irregularities, in the form of a peak followed by a dip at a higher frequency. Thus it is convenient to fill up the cavity if the tweeters are small enough to be mounted partly within the thickness of the front panel. It is also, of course, desirable to mount the tweeters as close as possible to the main loudspeakers so as to avoid any unnatural effects due to noticeable separation of the high- and the lower-frequency components of a sound source. Small tweeter units have sometimes been mounted in front of and within the cone cavity of the main moving-coil loudspeakers. The rear side of the tweeters has to be closed off from the sound radiation of the main cone, so as to make the tweeter a stiff sound source and to prevent it being driven acoustically at lower frequencies, with the result that it is difficult to make the unit small enough to avoid an appreciable obstacle in front of the main cone, which may cause irregularities in the frequency response. Co-axial mounting of a tweeter in this way can be successful if the unit is small enough to become effectively a part, or an extension of the main unit centre pole. Generally, however, it is easier to mount a tweeter separately within about 4 in. of the edge of the next lowest unit covering the frequency scale.

Fig 42 shows a modern ultra-high-frequency moving-coil tweeter which is designed to cover the frequency range from about 6 kHz to above the range of audibility. It exploits the use of a flat ceramic magnet which not only gives a high magnetic flux density but also achieves a very flat type of loudspeaker unit, which is thin enough from front to back to be accommodated partly within the cabinet front and thus to avoid any harmful cavity-resonant effects. In a typical design, the front may be between $\frac{3}{4}$ and 1 in. thick and the tweeter may be inset and mounted

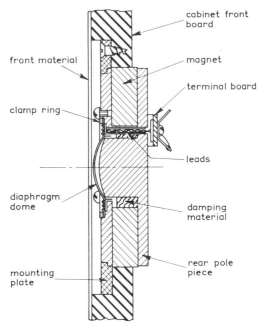

Fig 42. *A sectional view showing the construction of an ultra-high-frequency tweeter loudspeaker of the direct radiator moving-coil type. The response is maintained to over 20 kHz. This type of unit is used in the highest-quality loudspeaker systems to extend the range above that of normal tweeters and to ensure the best possible response to transient sounds.*

by means of the front flange so that the back is sealed and the diaphragm is clear of the cabinet-front covering material, as shown in Fig 42. The front covering material may be a semi-rigid material such as expanded metal or may be a stretched open-weave material of the Tygan type. Either of these materials may be supported on the cabinet front by a plastic foam flexible backing material cut away round the loudspeaker openings. Owing to its absorbent nature, this will not result in any increased cavity-resonant effects, and it will give an increased clearance between the dome of the tweeter diaphragm and the front covering material.

The diaphragms of these tweeters and those of somewhat larger units (which have a response which can be used lower in the frequency scale, eg down to 3500 Hz) are more akin to those of

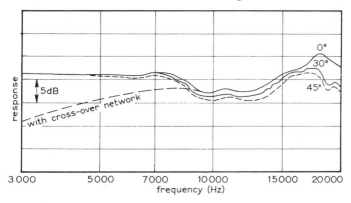

Fig 43. *The response of an ultra-high-frequency tweeter similar to that shown in Fig* 42.

moving-coil microphones than those of cone loudspeakers, in that a small diaphragm moulded from a thin rigid plastic material is used. The centre part of the diaphragm is domed outwards for rigidity, rather than made in the form of a concave cone, as in larger loudspeakers. The surround is relatively large and is given a controlled restoring compliance by means of tangential flutes or a flat radial roll section. The outside diameter of the complete

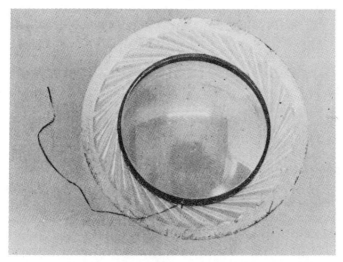

Fig 44. *The lightweight* 1-*in. diameter plastic diaphragm and moving-coil system of the ultra-high-frequency tweeter of Fig* 42.

151

diaphragm is usually not more than $1\frac{1}{2}$ in. so as to ensure freedom from break-up modes and a wide polar spread of high-frequency sound radiation (ie an approach to omnidirectional polar response). The speech coil is wound from copper or aluminium wire and is "formerless"; that is, it is wound as a self-supporting two- or four-layer coil cemented to the diaphragm. In the unit of Fig 42, some acoustic resistance damping below the coil slot is added in the form of a ring of porous plastic foam. The tweeter and its flange are preferably made a sufficiently good fit in the cabinet front so as to seal the enclosed loudspeaker cabinet and to avoid any air-leaks. A little putty or sealing compound may be added if necessary to ensure that the joint is sealed.

Loudspeaker cabinets or enclosures

The cabinet not only serves as an essential mounting and protecting frame for the loudspeaker units but also fulfils an important acoustical function. This is to control or to eliminate altogether the out-of-phase radiation from the rear of the loudspeaker diaphragm.[18,19,20,21]

Many different types of cabinet, baffle or enclosure have been evolved in order to give the desired results, but certain problems are common to all. Apart from the difficulty of obtaining a close approach to the ideal acoustic function at low frequencies, difficulties arise owing to mechanical resonances of the panels forming the walls and back, and also to acoustic resonances of the air enclosed within the cabinet. Both these types of resonance will be excited by the loudspeaker units, the mechanism being that energy is absorbed when the loudspeaker sounds the appropriate note; after the note has ceased, the energy stored in the mechanical and acoustical resonances of the cabinet are given back and radiated as sound, either by driving the loudspeaker cone from behind or by the continued vibration of the cabinet walls. The effects occur at low and middle frequencies and tend to cause a "coloration" of the sound reproduction, which is often described as "booming" or "honking," when it is particularly obtrusive. Obviously, it is very difficult to eliminate these effects completely and they are undoubtedly a factor which influences the character of the sound produced by particular loudspeakers and gives them their individuality. Generally speaking, middle-frequency resonances are the most objectionable but are more readily cured by using sufficiently thick and well-braced cabinet walls and by including a proper amount of cabinet internal acoustic damping in the form of sound-absorbing material.

Low-frequency mechanical vibrations of the cabinet walls are very difficult to eliminate completely, as it is not easy to make them thick enough to be completely rigid or so heavily damped as to be completely dead or inelastic. In fact, a certain amount of controlled cabinet-wall vibration may be accepted as having a sounding-board effect, as with many musical instruments, adding some fullness to the low-frequency reproduction. Nevertheless, the effect must be under control and the modern tendency is to make the walls as rigid and as dead as possible, by using suitable thicknesses of material, braced if necessary.

To a fair extent, the final performance demanded of a loudspeaker is in the realms of art rather than science, the choice often depending on individual taste, and often to a large extent independent of such factors as the cost and complexity to which one is prepared to go. One must, however, beware of certain traps when making choices based on limited initial listening tests. Some loudspeakers may sound impressive on certain programmes only or may have defects which become increasingly obtrusive in time and become annoying or even fatiguing to listen to. Excessive or boomy bass or hard peaky top may well come in these categories, as certain programmes may sound effective whilst many others may be unsatisfactory. However, it is noteworthy that nowadays the loudspeakers produced by well-established and experienced manufacturers are generally of good quality and have obviously proved acceptable to a large number of listeners with many different set-ups. Different programmes may still benefit from an intelligent use of the bass and top tone-controls provided on good amplifiers. Some suggested forms of listening test are given in the later section on sound-quality judgments.

We will now examine various types of cabinet and consider some examples of good modern design.

It is a fundamental tenet that loudspeaker enclosures have to be designed to suit the type of moving-coil unit to be used, particularly as regards the effective area and mass of the diaphragm and the centring stiffness. To some extent, the opposite applies, in that loudspeaker units may have to be chosen to suit the cabinet.

Tuned enclosures are mainly of the Helmholtz resonator or reflex-ported type. Quarter-wave tuned pipe or labyrinth types of cabinet have been used but tend to be more expensive to construct.

Ported reflex enclosures

In the reflex system the loudspeaker cone and suspension forms a series mechanical resonant circuit and the enclosured air volume forms a parallel resonance with the mass of the air associated with the port (actually the mass of the air in the port pipe plus the radiation mass of the air added to the orifice of the port). If the two resonances are tuned to the same frequency (ie if the unit bass resonance coincides with the frequency to which the cabinet resonator is tuned) the result is a coupled double resonant "circuit," exhibiting two resonant peaks. One is below and one above the bass-resonant frequency of the unit. If these two peaks can be damped to the correct degree, the bass response of the loudspeaker is extended considerably below the bass-resonant frequency of the unit. In practice it is quite difficult to obtain the correct damping of both the resonant peaks in the overall bass response unless a fairly large cabinet is used, with correctly positioned acoustic-resistance material mounted in exactly the right places. A common defect with reflex cabinets of compact size is that the upper bass-response peak is not well damped, producing an unfortunate bass "coloration" at a relatively high bass frequency (often 120–150 Hz).

Nevertheless, it is possible to make bass-reflex cabinet designs which work correctly, always provided that the proportions of the port, the resonator tuning and the damping are precisely right. Generally, expensive laboratory equipment is needed to achieve this result, though simple impedance measurements at the speech-coil terminals can be a good guide. The enclosure walls must be thick enough to be substantially non-resonant and internal damping material has to be correctly disposed inside the cabinet, firstly to damp the resonant upper peak of the two which occur and, secondly, to damp air-column mid-frequency resonances inside the box which may otherwise cause coloration. These internal resonances may also be radiated out through the port opening as well as driving the loudspeaker unit from behind. It is necessary for all joints in the cabinet and such places as loudspeaker-unit mounting flanges and terminal plates to be fairly air-tight, otherwise the port resonator tuning may be upset and sound leaks may occur.

A number of designs of ported cabinets are produced by the various loudspeaker-unit manufacturers, and the home constructor may obtain detailed plans from them. It is usually necessary, however, to follow the constructional details, dimensions and recommended materials fairly closely. In some cases, the proper damping materials and even complete properly damped port units are available from the manufacturers concerned.

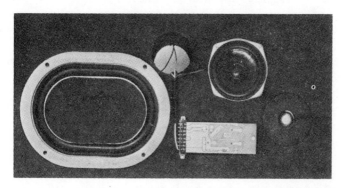

Fig 45. *A high-quality three-way loudspeaker module kit for assembly into home-constructed or ready-made cabinets to give a frequency range from 30 Hz to well over 20 kHz at a power rating of 30 watts. The large elliptical bass unit has an aluminium-reinforced expanded-polystyrene diaphragm and covers from 30 to 300 Hz. The 5 in. mid-range unit has a specially designed and treated diaphragm with a highly compliant linear-roll suspension. It covers the range from 300 Hz to 3 kHz here, though it may be used as a bass and mid-range unit in smaller enclosures. The ultra-high-frequency tweeter gives a response from 3 kHz to 30 kHz. The printed circuit three-way cross-over network is shown. The high- and mid-range units are isolated at the rear in their own enclosing cases and the low-frequency bass port opening is seen by the bass unit. Cabinet details are provided.*

An interesting variant to the port or open passage is to provide a subsidiary "slave"-driven diaphragm which is given optimum properties and may enable more compact cabinet proportions to be used.

Fig 45 shows a loudspeaker kit of high quality which is supplied with damping material and cabinet constructional details.

THE INFINITE-BAFFLE OR ENCLOSED CABINET

This is probably the most popular type of enclosure in use today. The enclosed volume of air acts as an "air spring" and adds directly to the total restoring stiffness of the cone, so specially designed units with low-stiffness surrounds and centring bellows have to be used with these cabinets if they are to be of reasonably small size.

The completely enclosed infinite-baffle box-type of enclosure offers some considerable advantages over most other types of cabinet as regards appearance, cheapness, compactness and performance, and a number of special loudspeaker units are produced for use in enclosed boxes. These designs now represent an advanced form of technique and have a wide frequency response and reduced non-linear distortion. As we have noted, the stiffness of the air in the enclosure constitutes an extra spring acting on the diaphragm. It adds directly to the stiffness of the surround and centring bellows. The low-frequency resonance between the total stiffness and the diaphragm mass represents the lowest bass frequency which the loudspeaker can radiate efficiently. This resonance without any added box stiffness usually occurred at about 40–50 Hz on the older types of high-quality loudspeaker unit. Adding a box enclosure of normal volume could raise the resonance and hence the bass cut-off frequency to about 100 Hz. In order to radiate bass efficiently down to 40 Hz, the loudspeaker resonance in free air has to be reduced to about 15 Hz. This means that very free surrounds and centring bellows are needed. The centring bellows must still ensure that the speech coil remains centralized in the radial magnetic gap under all conditions. It must also give a linear restoring force and must resist the onset of mechanical fatigue. Similar considerations apply to the surround. This must also form an air-tight seal and be properly damped so as to act as a correct terminating mechanical impedance to wave motion traversing the diaphragm outwards from the speech-coil. If this termination is not correct, a reflected wave will travel back along the diaphragm towards the coil and wave interference will occur at certain frequencies, resulting in an irregular response.

Generally a modern "infinite-baffle" loudspeaker unit has a good all-round performance and is characterized by a stiffer, more rigid and heavier diaphragm than before. The extra weight helps to keep the bass resonance low without making the centring bellows too weak. A relatively large and soft roll surround is often used, which is arranged to have a resistive nature so as to provide the right mechanical impedance and damping properties. The radial magnetic gap may be increased so as to allow more speech-coil clearance to compensate for the weaker centring action of the bellows. The heavier diaphragm behaves more nearly as a piston and has fewer severe break-up resonances than before, but the increased mass and the wider gap tend to reduce the efficiency. However, modern magnet materials such as ceramics or columnar cast magnets are capable of providing a high magnetic flux without any increase in weight, so that the efficiency can be maintained at a good level.

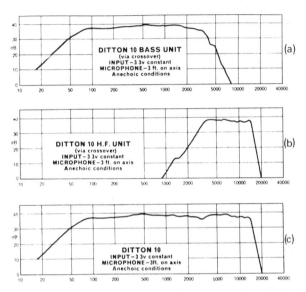

Fig 46. *The response curves of a modern compact bookshelf type high-quality loudspeaker.* (a) *The response of the bass and mid-range unit, which is a special 5 in. loudspeaker with a high-compliance cone suspension with a basic resonance of 30 Hz and capable of an excursion of* $\frac{1}{2}$ *in.* (b) *The response of the high-frequency moving-coil pressure direct radiating unit, which is* $1\frac{1}{2}$ *in. in diameter and is mounted close alongside the main unit in a compact enclosed cabinet.* (c) *The response of the two units when combined via a three-element cross-over network* (3·5 *kHz cross-over point*).

The result is a reduction in the size of the cabinet required to take even three-way loudspeaker systems using 12 in. woofer or bass units. A cost reduction, ease of housing, and a reduction in size of the cabinet bring other subsidiary benefits. Smaller cabinet wall panels are more rigid for a given thickness and hence do not require much damping or bracing; also acoustic edge diffraction effects are reduced in severity. Here the wave radiated from the diaphragm expands smoothly until the edges of the hemispherical wavefront meet the front edges of the cabinet; the pressure wavefront then suddenly has to diverge to propagate itself around the cabinet. The result is a discontinuity which, as always in wave propagation, causes diffraction irregularities to occur. The smaller the cabinet, the further up the frequency scale before the onset of severe diffraction effects is met. These effects cause measurable irregularities in the frequency response, but at higher frequencies the radiation from a diaphragm becomes concentrated in a narrower solid angle in the forward direction and cabinet-edge diffraction is then of small importance. Other precautions have to be taken in the design. The most prominent internal air-column resonances are the axial resonances between the three pairs of parallel walls. These resonances will occur at middle audio-frequencies in smaller enclosed cabinets and must be critically damped by draping or fixing suitable acoustic-absorbing material inside the cabinet. If undamped, these resonances will be excited by any corresponding frequency applied to the diaphragm and will affect the acoustic impedance loading on the rear of the diaphragm, causing an irregular frequency response. They may also persist after the exciting signal has died away, in which case, as we have seen, the diaphragm may be driven from the rear by the box-resonance to cause "honking" or resonant hangover. If the interior of this box is excessively damped by overfilling with absorbing material, then a large resistive load is acoustically coupled to the rear of the diaphragm which may then be over-damped, with a resultant loss of efficiency and a slowing-up of the rise time of transients. It is also desirable to maintain a nearly air-tight seal at all joints, screws etc. Effective sound radiation demands high power and the generation of very high sound-pressure levels in the enclosed box and air may escape at high velocity from even small leaks. Tweeter units mounted in the main cabinet must have their backs effectively closed off and must be well sealed into the cabinet round their periphery. It is desirable that the internal acoustic damping material used inside the cabinet should have the ability rapidly to take up the small variations in heat of the air which occur during the acoustic cycle. If there is no heat loss to the surroundings during an acoustic compression cycle occurring at low frequencies, then the process is classed as "adiabatic" and a maximum value of air stiffness will be provided by the air volume in question. It is well known that the work done in compressing air is stored as heat and thus the air temperature is raised. If this heat is transferred to a distributed material in contact with the air then on decompression less work will be returned from the air and the net effect will be that of a reduced

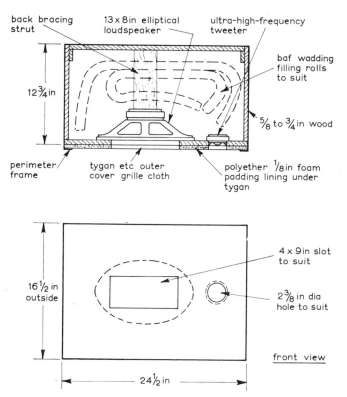

Fig 47. *An economical design by the author for a home-constructed two-way loudspeaker system. A wide-range elliptical paper-cone unit with a compliant surround is used in an enclosed cabinet with an ultra-high-frequency tweeter fed by a simple blocking capacitor. Good-quality* 13 × 8 in. *commercial elliptical wide-range units are available with low bass resonance suspensions. Suitable tweeters are now available from several leading makers.* (Units by Fane, ITT etc.)

stiffness. Materials such as BAF (bonded acetate fibre) wadding have been found to be good cabinet-damping materials for this reason, as well as being light and clean.

Infinite-baffle loudspeaker design thus requires that the

correct bass units are used and also that the construction and the electrical (amplifier) damping as well as the cabinet internal damping are correct. If all these factors are not properly balanced, the bass response may be peaked or even largely deficient.

Good examples of enclosed cabinet design are seen in the excellent performance obtained from "bookshelf" and "slim" loudspeakers now available using highly developed bass units capable of large linear diaphragm excursions, often combined with good tweeter units.

Unpacking and assembling loudspeaker units

Although modern loudspeakers are strongly made, it is important to take certain precautions when handling units and assembling them into cabinets—

(i) Never drop loudspeaker units even when packed, or subject them to any heavy air blast or suction, such as that which may result from slamming tightly fitting cupboard doors. A unit should not be abruptly placed or removed from a table with the cone face down.

(ii) Do not allow any ferrous metal objects such as tools, screws, scissors, needles etc to come near to a loudspeaker. Above all never place a loudspeaker on a work-bench where there is a chance of steel filings and swarf being attracted into the gap. Very fine swarf can penetrate dust-excluding gauze caps over loudspeaker gaps and it is impossible to remove unless the unit is demagnetized and taken apart—a job for the manufacturer.

(iii) It must be ensured that the cabinet front mounting surface is sufficiently smooth and flat to enable a good seal to be established without distorting the loudspeaker chassis or mounting ring by undue tightening of the securing screws. It is advisable to use brass or bronze screws in order to avoid the possibility of magnetic slivers being produced. The screws must be strong enough and long enough to support the unit. Any shocks during transit can generate large acceleration forces on heavy loudspeaker units.

Testing and quality judgments on loudspeakers

A number of important tests may be made on loudspeakers in order to eliminate defects and also to make satisfactory comparisons and judgments of quality. The tests proposed here may readily be carried out without the use of special test equipment.

1 Continuous tone testing. A large low-frequency tone may be obtained by such devices as placing a loop of insulated wire from the input of the amplifier near to a mains transformer in the power pack. It will give an idea of the bass capabilities and will tend to show up defects such as buzzes and rattles in the unit or cabinet, as well as checking on correct phasing of dual bass units. A phase reversal shows up as a drop in bass level. However, it is necessary to avoid applying constant tones to loudspeakers at a high level as there is a risk that the speech-coil may be over-heated and burnt. Normal programme peaks are of such small duration that speech-coil heating does not occur, except possibly from sources such as electric guitars or electronic organs, which demand heavy-duty loudspeakers.

Buzzes in loudspeakers may be due to loose washers or terminals etc, or may be due to loose front fabrics or expanded metal. Mounting these tightly over a foam underlay (with cut-outs for the loudspeaker openings in the foam) will often prevent these troubles.

2 A useful continuous random tone for testing is often readily obtained by tuning an FM set away from any station, if necessary removing the "squelch" noise-suppressing circuits. The resulting audio signal is close to "white noise," so called by analogy with white light, which contains all frequencies in the visual spectrum at equal intensities. If the lower frequencies are relatively increased in level by means of the tone-control bass boost, the result is "pink noise," which is sometimes claimed to be closer to the frequency distribution in normal programmes.

Listening to random noise on loudspeakers can give quite a lot of useful information. The higher frequencies should be well audible as a smooth rushing sound. Acoustic colorations or transient hangover tend to show up in the form of alterations in the evenness of the random noise. Listening off the axis of the loudspeaker gives a good idea of the distribution of high frequencies in the room.

3 Speech can be a good test of accurate reproduction, especially if the original recorded talker is able to speak in the same position as the loudspeaker in the room. However, male speech

in particular must be reproduced at the same level as the original or the low frequencies will be exaggerated according to the low-frequency auditory contours of the ear if the level of reproduction is too high. Sibilants may also be exaggerated in some cases.

4 When music is used as a test, apart from certain specially recorded passages on test records which are designed to show up certain effects, it is advisable to use well-balanced musical selections free from over-modulation which might give rise to tracing distortion on gramophone records, where these are used as a programme source. It is best to avoid any recordings in which special effects may have been used as these may prejudice the judgment of quality.

A tendency for all bass notes to sound muddled and indistinguishable can indicate resonant bass coloration and intermodulation. Piano music is a good test of quality. Poor transient response, any appreciable intermodulation and wow or flutter will all be shown up. If string tone is correct and if castanets, cymbals and the triangle stand out sharply and with good "presence," an extended top response is indicated.

In general, all reproduction should be improved by stereo. A sense of space and another dimension are added to the reproduction and the so-called monophonic "hole in the wall" effect is largely removed. The overall result should be more natural and presence should be improved. The effects of intermodulation and other imperfections are usually less noticeable and listening generally is less fatiguing and more satisfying. The spatial stability and apparent direction of individual instruments and soloists should be reasonably good and a sense of spaciousness should be associated with the reverberant sound as well as with the direct sound. On stereo, one is checking the audible effects which depend on channel matching, "cross-talk," loudspeaker characteristics and positioning, in addition to the general parameters of reproduction which also apply to monophonic reproduction. Dramatic pieces involving rapid movement and action can have an increased emotional impact when reproduced on a good stereo system. Any appreciable interchannel level mis-match or cross-talk may considerably reduce the dramatic appeal and sense of rapid transition involved in a programme.

8

Advanced loudspeaker systems

It is generally conceded nowadays that, except for certain special-ized fairly low-power units, high-quality loudspeakers of the moving-coil type require more than one drive unit in order to cover the frequency range effectively. There are several reasons for this, some of which have been outlined in the previous sections on loudspeakers. It is difficult to design a cone or diaphragm system to cover a wide frequency range without irregularities in response due to break-up resonances and standing-wave reflections from the cone edge. Non-linear "Doppler" distortion due to handling high and low frequencies simultaneously, difficulties in maintaining a uniform polar distribution or directional response, as well as various high cost factors involved in the manufacture of efficient wide-range units, all must be considered.

One might imagine that membrane-type loudspeakers of the electrostatic type or the "printed circuit" electro-dynamic mem-brane type could easily provide single-unit wide-range loud-speakers. However, examination of high-quality designs of this type shows that it is necessary to sub-divide the membrane into separate high- and low-frequency areas of different sizes so as to maintain a reasonably constant polar response. The situation is thus broadly similar to that of moving-coil units, where we have to divide the frequency range electrically into suitable channels for feeding the bass, mid-range and high-frequency units via care-fully-designed cross-over networks, which are usually fed from the common power amplifier output for the programme channel in question.

A wide frequency range, extending up to the limits of audibility, is necessary for the true reproduction of all sounds, especially transients. This is available from the improved stereophonic

VHF broadcasts now obtained as a result of the use of high-quality pulse-code-modulation music line transmission links and other improved equipment between studio programme originating points and the transmitters. The widespread use of noise reduction techniques of the Dolby type and the use of improved magnetic-tape masters have extended the range of gramophone discs. Magnetic tape and cassette recordings are also now improved.

For some while, good grade amplifiers, gramophone pick-up cartridges and other parts of the reproduction chain have been capable of handling the available extended high-frequency response, and hence it is now more than ever worthwhile to examine means of improving our loudspeaker systems, bearing in mind that stereo demands up to four high-quality channels for the modern "surround sound" or the true quadraphonic pro-grammes which are now becoming popular.

The best commercial multi-unit loudspeakers have always necessarily been expensive, particularly when housed in the massive carefully made cabinets designed for professional monitoring purposes. It is thus very useful for the keen amateur to see how far he can help himself by taking advantage of the various high-quality kits and basic loudspeaker assemblies which are now available, and also to see how these can be housed or built-in to the home. Many manufacturers now give elaborate cabinet constructional details.

In the last few years the better grades of loudspeaker have been refined and improved to such an extent that it seems unlikely that any very marked progress will be made for some while. This being so, users may well feel that it is worthwhile investing in really top-class loudspeakers which are likely to give satisfaction from high-quality programmes for some time to come. All good-quality loudspeakers in cabinets tend to be expensive, and thus it may be worth the extra expenditure to obtain loudspeakers which are the best available, and which are likely to remain a source of pleasure and satisfaction for many years to come. Nothing is more dis-appointing than to expend a fair amount of money and then, a few months later, to come to the conclusion that something better is required in order to do full justice to the programmes you have available. In these days of quadraphonic sound, it is even more important not to make such mistakes!

Siting and mounting arrangements for loudspeakers

It will be appreciated that the positioning and angling of directional loudspeakers can be quite important if one is to obtain the best stereo effects and frequency coverage over the desired listening area in the room.

Floor-standing may not always be desirable, except for loudspeaker cabinets of a fair height in which the tweeters are raised to near ear level for normal seated listeners. Floor-mounted loudspeakers may also be coupled to normal wooden suspended floors, thus exciting structural resonances and even disturbing pick-up tracking if the turntable cabinet is also on the floor.

Shelf-mounting is not always convenient or correct, particularly for larger loudspeakers. Some very useful adjustable wall-mounting plates are now available. These require secure screw-fixing to solid walls and are available to take loudspeakers of almost any size, these being firmly fixed and angled as required. A good example is the range from Hi-Fi Aids Ltd, Reading.

A case may be made out for mounting rear quadraphonic loudspeakers or "surround sound" radiators at a height considerably above that of the frontal stereo units, thus aiding three-dimensional ambient effects and avoiding disturbance of the frontal sound image. Angled wall mounts are ideal for this purpose and can also have the advantage of allowing floor space beneath for other pieces of furniture etc.

Fig 48 shows the basic corner placement of loudspeakers for quadraphony in a rectangular room plan. There is considerable scope for variation in the siting and the angling of such loudspeakers according to the room arrangement, the type of loudspeakers and the listening area desired.

It is often thought that the actual number and disposition of the loudspeakers must be directly related to the number of channels used. Thus, two loudspeakers for frontal stereo, four loudspeakers (in the room corners) for quadraphony. This is not necessarily the case, as there is considerable scope for experiment in both the number and the spatial disposition of the loudspeakers for any given system. In other words, two-channel frontal stereo signals feed left and right loudspeakers, which may, if desired, be "doubled up" in order to spread the respective signals and possibly to improve the effect by avoiding "point sources." More particularly, rear or side signals in quadraphony may be fed to any desired

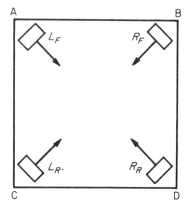

Fig 48. *A possible arrangement of front and rear loudspeakers for quadraphonic reproduction. The effective listening area in the room is enlarged compared to two-channel stereo and extends further towards the walls. Results are still influenced by the angling of directional or semi-directional loudspeakers and on the extent to which they are sited near to walls or corners. The same applies to omnidirectional loudspeakers.*

number of loudspeakers along the sides or back of the listening room. Again, a measure of improvement in the ambience may be obtained by feeding derived "pseudo-stereophonic" signals to a single rear loudspeaker or to a number of loudspeakers in parallel electrically, but physically placed along the rear and side walls. Some might even be placed near to the floor or the ceiling, in order to give a sense of height to the ambient sound.

Directional and omnidirectional loudspeaker arrangements

When specifying loudspeakers for stereophonic sound listening in an average room there is now a considerable choice, not only in the power output and response available from different types of more expensive loudspeakers, but also in the type of sound distribution which is provided.

It is fairly well established that loudspeakers with a directional polar response curve shape of the general form shown in Fig 3, p. 17, are capable of giving firm stereo images and a good "fusion" of sound to establish fairly precise virtual sound sources for positions intermediate between the loudspeakers. The central room listening area over which an accurate stereo picture is maintained may be enlarged somewhat by expedients such as angling the loudspeaker axes so that they cross towards the front of the listening position, but the available seating positions usually remain somewhat critical if accurate image location is desired. It is, of course, most desirable that the polar response shape should not change appreciably over the frequency range, particularly at high frequencies. Even the most advanced loudspeakers can only make an approximation to this requirement.

However, it is often argued that for the great majority of stereo programmes one is not really much concerned as to the precise position of sound images, so long as there are not any disturbingly unnatural effects, such as "disembodied high frequencies" or excessively "smeared out" sources such as, for example, a piano apparently occupying the full width of the stage. One does not want to know exactly where each member of an orchestra is seated; one is concerned with the general blend of sound, together with the spatial positioning and separation of the main portions of the orchestra. One is not conscious of the directions in space of the background reverberant ambience, though we find that this can be improved with the use of extra loudspeakers, quadraphony etc.

This being so, it has been found more recently that many people prefer to use loudspeakers which are more omnidirectional.

Fig 56 shows how the polar response may be progressively spread from the limited frontal angle of radiation produced by a directional loudspeaker out to the 360° coverage afforded by a completely omnidirectional system. Some loudspeakers achieve a spread of direction by expedients such as facing units upwards or sideways on to diffusing or reflecting surfaces. Unless this is carried out very carefully, unwanted phase shifts may be introduced by the differing path lengths which the sound traverses at different frequencies. The trend is now to provide more units facing outwards at different angles.

Forward-facing loudspeaker systems

Loudspeakers with specially designed directional characteristics are well known. Such specially designed units as horns, line sources etc, are used when it is desired to radiate sound within a restricted solid angle, thereby saving power and avoiding undesired sound reflections in auditoria and large rooms. Domestically, they have been used in the form of more precisely controlled units having a polar curve near to a "figure of eight" bidirectional curve shape.

Carefully-designed bidirectional membrane loudspeakers, such as the renowned Quad electrostatic loudspeaker, have a polar curve of this type, which is carefully maintained in shape up to high frequencies, and it is to be expected that very accurate stereo results will be achieved in average rooms.

Most loudspeakers of the enclosed cabinet type have pressure-type radiating surfaces, with the result that they are inevitably virtually omnidirectional at low frequencies, but become directional at higher frequencies, due to the normal interference focusing effect of a diaphragm for the shorter wavelengths of sound at high frequencies. Careful design of the tweeters is still necessary so as to

Fig 49. *The Ditton* 66 *and Ditton* 25 *multi-unit high-quality forward-facing cabinet loudspeakers. The* 12-in. *bass unit is aided at very low frequencies by an acoustically-driven slave diaphragm* (Rola Celestion, Thames Ditton).

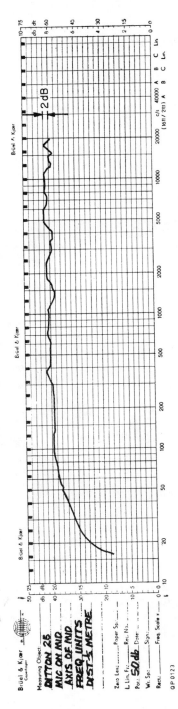

Fig 50. *The response of the Ditton 25 cabinet loudspeaker shown in*
Fig 49 (Rola Celestion, Thames Ditton).

prevent the polar response becoming too directional in the upper ranges.

The overall result is found to be a satisfactory compromise for a great many listeners, because an omnidirectional response at low frequencies in most home-listening rooms helps to maintain a good bass response and does not appreciably diminish the accuracy of stereo image positioning, as this is largely governed by the shape of the polar curve at higher frequencies. In fact, most high-quality monitoring loudspeakers are designed in this way.

Figs 49, 50 and 51 show typical examples of high-grade cabinet loudspeakers with front-facing units and the responses achieved.

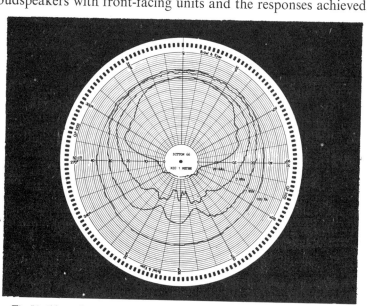

Fig 51. The polar response of the Ditton 66 cabinet loudspeaker. Careful matching of the mid-range and high-frequency tweeter response and grading of diaphragm sizes gives very even coverage of high frequencies up to 20 kHz over a wide frontal angle (Rola Celestion, Thames Ditton).

Although these would not normally be particularly classed as directional loudspeakers, it is reasonable to consider them as such, compared to loudspeakers where multi-facing units and other expedients are adopted to make the response much more strictly omnidirectional at middle and high frequencies.

An important point is that much attention has to be paid to achieving a smooth cross-over between woofer or mid-range units and tweeters, both for the frequency response and for the polar response. This not only means that the transient response and the relative phase between the two types of unit must be correct in the cross-over region where both are radiating sound, but there must also be a good match in the polar response shape of the units. If this is not so, there may be a marked step in the frequency response for a listener who subtends an angle of 30° or more off the main axis of the loudspeaker. It is obviously more difficult to meet these requirements when separate units are involved, even when they are mounted closely side-by-side. One solution is the use of concentrically-mounted tweeters, though care must be taken that the obstacle effect of the tweeter does not interfere with the response of the woofer. Fig 52 shows a design where a co-axial horn-type tweeter is built within the centre pole of the woofer, thereby avoiding this difficulty.

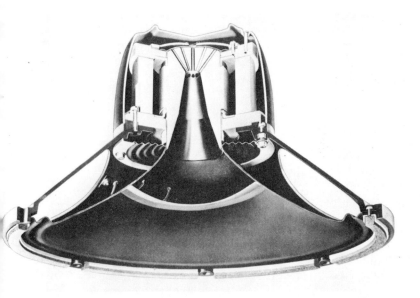

Fig 52. *The Tannoy dual-unit loudspeaker in which the wide-range tweeter is assembled on the rear of the woofer and shares the same magnetic pole system. The multiple bores in the centre pole act as an efficient throat coupler to the tweeter diaphragm, the curved woofer cone acting as a flared horn. The concentric mounting avoids difficulties due to phase displacement which may occur with separately mounted units* (Tannoy Ltd, London).

The low-frequency loudspeaker

We know that a major problem is met in obtaining a really good bass response for the high-quality higher-power loudspeakers which are needed for the best sound reproduction. In order to radiate the necessary sound power at the lowest bass frequencies (below about 70 Hz) a fairly large diaphragm is required if excessive displacement and some inevitable accompanying non-linearity are to be avoided. The use of larger diaphragms above about 8 in. (20 cm) diameter means that large and rigid infinite baffle (enclosed) cabinets or large-reflex enclosures often have to be used, particularly if the very lowest frequencies are to be radiated effectively.

In these cases there may be considerable advantages in other forms of enclosure design, such as a folded horn construction or labyrinth rear enclosures. These forms of construction are expensive to use in commercial designs because of the relatively large amount of carpentry and cabinet making involved. This, however, is an area where the amateur with moderate woodworking skills and facilities may be able to achieve excellent results, particularly as some of the best loudspeaker manufacturers now sell loudspeaker-unit kits and matched cross-over networks with instructions for the home construction of advanced cabinets of the more elaborate types.

Following on from the earlier description of enclosed and reflex types of loudspeaker, we may now consider some more advanced designs intended to give wider frequency range and better sound distribution, due largely to the more effective utilization of the loudspeaker units.

Labyrinth loudspeaker cabinets

This principle was evolved some years ago as a logical solution to the problem of absorbing the anti-phase low- and middle-frequency sound radiated from the rear of loudspeaker cones. In the totally enclosed cabinet a fairly large volume of air is coupled to the cone. The volume has to be large enough not to add an excessive amount of air stiffness to the cone and so raise the effective lower resonance, which determines the bass cut-off frequency. High-compliance units with flexible surrounds and sometimes high-mass cones are used to keep the resonance at a low frequency without an altogether excessive cabinet volume. The presence of this air volume also means that middle-frequency air-volume resonances occur at

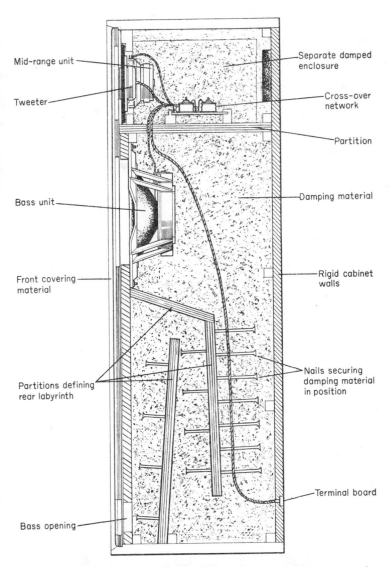

Mid-range unit

Tweeter

Bass unit

Front covering
material

Partitions defining
rear labyrinth

Bass opening

Separate damped
enclosure

Cross-over
network

Partition

Damping material

Rigid cabinet
walls

Nails securing
damping material
in position

Terminal board

Fig 53. This shows a damped labyrinth or acoustic line treatment housed as a folded pipe within the cabinet. Lightly packed damping material is distributed throughout the length of the pipe and held by wires or nails. The length is tuned so as to give an augmented output at very low frequencies, higher frequencies being completely damped out. An open-backed mid-range unit and a closed-back tweeter are housed in an upper entirely separate enclosure, optimum damping being applied to the mid-range unit in order to achieve a smooth roll-off (Radford Acoustics, Bristol).

harmonics of frequencies given by 550/*d* Hz, where *d* is the internal distance in feet between cabinet walls. Thus for a cabinet of internal size 1½ ft (0·5 metre approximately) air resonances will occur at frequencies of about 350 Hz and upwards. Unless sufficient internal acoustic damping material such as wool, BAF wadding etc, is added, these middle frequencies may well become audible as middle-frequency tone coloration in the form of a hangover or "honking" effect. Attempts to add enough damping may often result in the main resonance of the loudspeaker being over-damped thus lowering the whole bass response and leading to the negation of the whole exercise, which was to obtain a full bass response!

Similar difficulties apply to reflex-cabinet designs, with the added complication that the tuning of the cabinet and the port may be upset by adding internal damping.

A labyrinth represents a folded damped pipe which is closely coupled to the rear of the diaphragm, with a minimum volume of air left between the loudspeaker unit and the start of the labyrinth. If the labyrinth can be made of the correct area and is correctly damped along its length, it can be made to terminate the diaphragm in its "characteristic resistance." This is the value of acoustic resistance which would completely absorb a sound wave launched into it, without any reflections or standing-wave effects.

In practice, complete absorption means the use of a rather long pipe with acoustic damping material distributed at fairly low density along the length. It was discovered that if the labyrinth

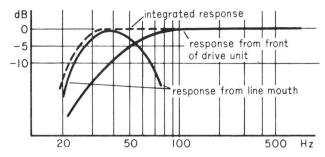

Fig 54. *The frontal sound radiated from a well-damped woofer will tend to fall off below* 100 *Hz, as shown. The response from the mouth of a well-designed rear acoustic line or labyrinth can be tuned to respond in phase with the output from the cone at frequencies below* 100 *Hz, the overall result being a smoothly maintained bass response down to below audibility* (Radford Acoustics).

pipe was made of such a length that it was tuned as a quarter-wave resonator for the lowest frequency of interest, the far end would emit sound which assisted radiation at the point where that from the diaphragm was falling off. All sound in the labyrinth may be effectively absorbed at frequencies above the resonance by means of distributed damping of the correct grading.

In practice the pipe is achieved without too much complication by fixing partitions into the cabinet in the manner shown in Fig 53.

Fig 54 shows an ideal response obtained at the lower end of the bass by this means.

Omnidirectional loudspeakers

It appears that more and more people appreciate the improved spatial and stereo ambient effects which accrue from the use of these loudspeakers, and we see that they are becoming available quite widely in the more expensive ranges offered by some makers of high-quality loudspeakers. The extra number of loudspeaker units used necessarily increases the cost, but in addition to a much wider angle of radiation and a less critical listening position, the distortion level is further reduced as the power output is spread among more loudspeakers in the critical middle- and upper-frequency bands.

These loudspeakers naturally tend to be large and are usually housed in individual floor-standing cabinets, without other pieces of furniture etc placed in the near vicinity. Hence they are more suited to the larger listening rooms which are preferred for quadraphony.

At least one manufacturer offers the units, dividing networks and cabinet constructional details for home construction of high-quality omnidirectional loudspeakers with various numbers of units angled to face in various desired directions (Radford Acoustics, Bristol).[26]

Omnidirectional loudspeakers, particularly those required to handle larger power, may use several mid-range loudspeakers, often arranged to face in different directions. Fig 55 shows that an undesirable middle-frequency resonance may result if a restricted rear enclosure is used with mid-range units. The correct smooth roll-off may be assisted by the use of a larger correctly-damped enclosure within the main loudspeaker cabinet, open-backed mid-range units then being used rather than enclosed types.

An approach to an omnidirectional response may be achieved by simpler means than the use of multi-facing units. In designs such as that shown in Fig 66, page 207, a single wide-range unit is faced vertically upwards on to a curved diffusing surface. Limitations must exist in this simple approach in that the spreading of sound by such a surface is frequency-dependent and also standing-wave effects between the cone and the diffuser may cause irregularities in the response at high frequencies.

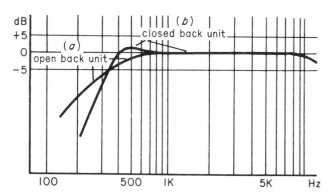

Fig 55. The response of a mid-range loudspeaker unit (a) with back of unit open or vented into a large damped cabinet volume and (b) with the unit back closed off by a restricted rear volume, which is common practice to avoid interference from the woofer. The effect is often to cause a middle-frequency resonance at about 400 Hz which may cause coloration. Any such enclosure must thus be carefully damped with absorbing material or, preferably, the mid-range unit is given a larger enclosure of its own (Radford Acoustics).

Fig 56 shows how an omnidirectional response may be approached in stages by increasing the number and the angles of units in steps, in order to increase the effective angle of radiation. A strictly omnidirectional response requires units to be faced in vertical directions as well as horizontal. However, this is more difficult to achieve than the horizontal spreading of sound by units arranged in one plane. For most practical purposes, it is sufficient to concentrate on dispersion in the horizontal plane, assuming that this coincides approximately with the height of listeners' ears.

Fig 57 shows the construction of an ambitious omnidirectional loudspeaker of this type for which unit kits and constructional plans are available.

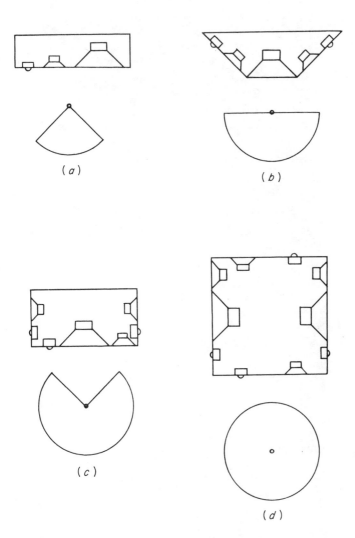

Fig 56. Diagrams showing the progressive stages by which an approach to true omnidirectional response may be obtained at middle and high frequencies: (a) shows that forward-facing units radiate sound into a restricted frontal angle at frequencies where diaphragm sizes begin to cause focusing of sound; (b) shows an improvement by angling the extra mid-range and tweeter units outwards; (c) and (d) show the progressive improvement as further units are added to face sideways and to the rear. The use of additional units is preferred to spreading sound by means of convex reflecting surfaces because these may introduce phase delays and standing-wave effects (Radford Acoustics).

Fig 57. *An advanced design of wide-range high-quality omnidirectional cabinet loudspeaker in which mid-range and high-frequency units are faced in many directions. The bass section uses a damped labyrinth termination similar to that shown in Fig* 53 (Radford Acoustics, Bristol).

Horn-coupled loudspeakers

The use of horns of expanding sectional area, to increase the matching efficiency of a loudspeaker diaphragm to the air, is well-known. The potential benefits are very considerable but there are various difficulties, the principal ones being:

1 Although an exponentially-expanding horn may give theoretically perfect loading of the loudspeaker unit, in practice there is an inevitable mis-match at each end, causing standing waves and a slightly irregular response unless the design is carefully modified in various ways.
2 An efficient low-frequency horn is inconveniently long unless it is folded. This generally precludes the simultaneous use of the horn for higher frequencies, because these are liable to be excessively reflected at the bends, causing standing waves and losses.
3 A horn with a comparatively large mouth-size tends to produce undue focusing of high frequencies.
4 It is difficult to make the horn walls sufficiently dead and non-resonant mechanically to avoid coloration of the response.

The advantages of horn loading are such that it is still worth while using the principle in specialized loudspeakers where cost is not particularly important compared to a high conversion efficiency, good damping of the diaphragm by the air load etc. Thus horns have been used in cinemas and large auditoria where efficiency and high-power output with low distortion are important.

It is usually necessary to design a horn for a restricted frequency range. Thus folded horns are used for low frequencies, while direct horns are used at higher frequencies, with special throat-coupling elements to avoid cavity effects where the horn meets the diaphragm, and features like cellular sub-divisions, acoustic lens-coupling plates or narrow rectangular mouth-aspect ratios. Domestically, the bass horn has long been featured in many successful loudspeaker designs. It solves the problem of obtaining good bass loading on a small diaphragm, which does not have to move very far to radiate considerable power at low frequencies.[28]

Fig 58 shows such a design, where higher frequencies are radiated directly from the front of the diaphragm, while the rear is coupled into a folded horn to give good bass loading. One advanced modern design of loudspeaker uses three units to cover the frequency range, each coupled to its own horn system, designed for maximum quality and efficiency (Lecson, St Ives, Huntingdon).

Thus the use of horn loading still has its supporters, though it is necessarily confined to the more expensive market for luxury loudspeakers, if high performance is to be achieved. One example is Lowther Acoustics Ltd, Bromley, Kent, who have produced high-quality horn-loading enclosures for many years, using high magnetic flux wide-range moving-coil driver units with folded-horn bass loading.

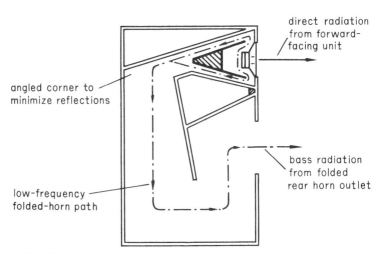

direct radiation from forward-facing unit

angled corner to minimize reflections

bass radiation from folded rear horn outlet

low-frequency folded-horn path

Fig 58. One method of obtaining improved efficiency of bass radiation. This shows a folded-horn system fed from the rear of a bass loudspeaker unit. The area of the horn is expanded in an approximately optimum manner by building internal partitions to form a throat-coupling unit and guiding walls within the cabinet. Reflecting surfaces at the bends may have some absorbing treatment to minimize reflection troubles. Crossover to the bass horn is at about 400 Hz, aided by the loss in higher frequencies in the throat behind the bass unit and by further losses at the bends. Some designs also include a shorter straight flared horn coupled to the front of the main unit, which is then a wide-range small type with a good bass response.

In some cases it may be found that the radiation of extreme bass is helped by standing the units in corners of the room, the diverging angles formed by the walls and floor acting as an extension of the horn expansion (Klipsch loudspeaker). Published designs for home construction of bass horns have been arranged for permanently building into room corners. Commercial units are usually necessarily free-standing cabinets and give more freedom for the user to find optimum positions in the room.

An interesting point in some designs is that the loudspeaker unit is coupled into the horn column at a distance effectively $\frac{1}{3}$ of the total length. This was long ago found to reduce objectionable odd harmonics of the residual standing-wave resonances which are liable to occur to some extent, because the horn is not in practice perfectly terminated at the mouth by the air load and thus has some organ pipe-type resonant effects.

Cross-over problem

It has always been appreciated that when the frequency spectrum is divided among a number of loudspeakers, some form of filter or cross-over system has to be provided in order to feed each unit with the appropriate part of the spectrum.

Originally this was approached by applying standard electrical filter-design concepts and evolving suitable high-pass and low-pass sections using inductance and capacitance. However, a number of difficulties are met:

1 Loudspeaker voice-coil impedances are low compared to the usual communication circuits in which filters are used, and the amounts of power are generally much greater when loudspeakers are driven directly.
2 The impedance of loudspeakers varies widely over the frequency range and the inductive and capacitive parts of the impedance may change rapidly in the region of loudspeaker resonances, thus completely upsetting the working conditions of many types of filter.
3 The delaying effect of the filter on transient wave-fronts and wave-phase changes may be important.
4 The network may be required to equalize the individual loudspeaker responses in order to eliminate resonances and to match up the units if their efficiency is at all different.

Cross-over network design

This formidable list of requirements has led to the evolution of quite complicated and expensive cross-over networks for the high-quality monitoring loudspeakers used for broadcasting and recording purposes. These are in fact necessary in many cases but, equally, it is often possible to obtain good results with relatively simple cross-over arrangements, provided that care is taken in the design and testing of these.

Theoretically, half the power should be passed to each loud-speaker unit at the cross-over frequency. Also, it is important that excessive phase shift should be avoided because this can adversely affect the transient response of the combination. Unless you have pulse-testing equipment, it is best to keep cross-over networks as simple as possible. These may be adjusted for phase by varying component values and if necessary moving the tweeter back and forth to find if its position relative to the other units is critical.

The actual cross-over frequencies depend on the units used. It is often desirable to cross-over at about 1 kHz from the woofer, so as to avoid break-up resonances of the larger diaphragm. Mid-range units usually respond well up to between 5 kHz and 7 kHz, where one can cross-over to the tweeter without fear of overloading it with excessively low frequencies. Super-high-frequency tweeters may cross-over at about 10 kHz and respond up to over 20 kHz.

The coils used in cross-over networks have to be designed generously, so as to avoid magnetic saturation and to have low resistance, in order to avoid power loss. They are usually wound as air coils of generous size in the more expensive loudspeaker combinations.

Simple single-element cross-over networks, which consist of a capacitor in series with the tweeter and an inductance coil in series with the low-range loudspeaker, will normally have a nominal roll-off of 6 dB/octave at cross-over (as used in the loudspeaker design on page 159). These are simple to make and adjust but may not reduce the input sufficiently outside the desired passbands, thus possibly overloading tweeters with lower frequencies and also spoiling the overall polar response. Sharper rates of cut-off of 12–18 dB/octave require multi-element networks and introduce greater possibility of phase errors and transient-response distortion, particularly at 18 dB/octave, which is near to the practical limit for most designs. Specially tuned and damped dip circuits may be introduced in particular loudspeakers to correct resonances or to remove troublesome frequency bands, but these again must be carefully examined for unwanted transient effects. Another useful function is to be able to introduce relative loss or gain between the various units in order to match them up for sensitivity, which often differs between tweeters and woofers. Resistive attenuators or loss pads may lose too much efficiency and require excessive amplifier power, and thus tapped transformers are sometimes used in cross-over networks on the more expensive loudspeakers.

Loss pads may be suitable for reducing the response of tweeters, where these are used with less efficient mid-range or woofer units, as there is not much loss of amplifier power involved. The converse case where the tweeters are less efficient is more serious from the loss point of view.

Other difficulties exist in that:

1 Cross-over networks and loss pads reduce the beneficial damping effect which modern low-source impedance amplifiers have on loudspeakers.
2 Loudspeaker voice-coil impedances vary widely with frequency (3/1 or more), which makes it difficult to optimize cross-over filter values.
3 Cross-over network reactive components may resonate with loudspeaker reactive speech-coil impedances at some frequencies, thus upsetting power-amplifier loading, with the risk of premature overloading and distortion at a now reduced output power level.

Active cross-over networks

It will be seen from the above considerations that cross-over networks of the more complex type are liable to be both expensive and difficult to design, and also difficult to adjust. As transistor amplifier modules are now becoming relatively cheap, it is an obvious move to use a separate power-amplifier to feed each loudspeaker unit, the filter characteristics being incorporated in the amplifiers as required. Generally, tweeters do not need very powerful amplifiers. This approach is now being followed in some high-quality monitoring loudspeakers.

It has now been shown that 12 dB/octave low-pass and high-pass filters, very suitable for building into active cross-over amplifiers, may be made by adding a few resistors and capacitors to standard type 741 operational IC amplifiers. These elements are then incorporated in preamplifiers driving the appropriate power-amplifiers. Figs 59 and 60 show the circuit modules (see *Wireless World* article by D. C. Read, December 1973). There is obviously scope for varying the relative gain and the rate of roll-off, as well as introducing other forms of desired equalization in this approach. The bulk and cost of low-loss cross-over coils is also saved, the cost of standard ICs and amplifier modules now being very reasonable. Multiple op amp ICs offer simplified filters (e.g. Siliconix).

The objections listed in the previous section on passive cross-over networks are almost all removed by the use of active systems,

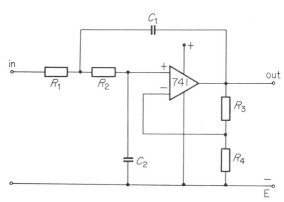

Fig 59. *Low-pass cross-over circuit element based on a standard IC operational amplifier. 12 dB/octave roll-off is produced when the circuit is inserted between amplifier stages. Cross-over frequency* $= 1/2\pi \sqrt{R_1 R_2 C_1 C_2}$ $R_3 \doteq R_4$ *for approximately unity stage gain* (D. C. *Read and* Wireless World).

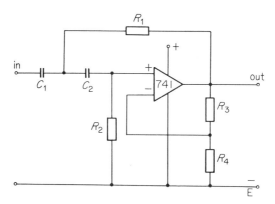

Fig. 60. *High-pass circuit similar to Fig 59 giving 12 dB/octave roll-off. Cross-over frequency* $= 1/2\pi \sqrt{C_1 C_2 R_1 R_2}$. *Band-pass cross-over characteristics for mid-range units may be obtained by using Figs 59 and 60 in successive stages* (D. C. *Read and* Wireless World).

as the separate power-amplifiers drive their own loudspeakers directly. The reduced average power requirements for tweeters means that heat sink requirements may be reduced when standard power-amplifier modules are used, thus saving weight and cost.

The above author also showed how 6 dB and 12 dB/octave stages may be used in successive preamplifier stages in order to achieve a "soft roll-off" cross-over characteristic free from ringing and transient distortion, but with the benefits of an 18 dB/octave or greater ultimate cut-off. This reduces any tendency to overload tweeters with lower frequencies and greatly dilutes the possible effects of out-of-band resonances which may be present in woofers and mid-range units.

Band-pass feeds for mid-range units are readily produced by cascading suitable high-pass and low-pass active filter stages. The required circuit values for the various filters may be calculated from the data given in articles, books, and by semiconductor manufacturers. It is obviously an advantage for constructors to check results with an oscillator, frequency sweep records or tapes. Test equipment of this sort may be acquired or built at reasonable cost by constructors and will prove a good investment, particularly if used in conjunction with a simple type of oscilloscope.

Enough has been said to show the way in which active cross-over techniques are likely to be used to improve the performance of multi-unit loudspeakers by precise control of the circuits and operating parameters.

Motional feedback control

It has been known for a long while that a very effective means of improving the response and the non-linear distortion of a loud-speaker is to obtain a voltage or current which is a perfect copy of the motion of the diaphragm or radiating surface and to incorporate this voltage within the negative feedback loop of the power-amplifier which drives the loudspeaker. This method is already used to control the response of moving-coil cutter heads for making master gramophone discs, and for large moving-coil vibrators used in testing aircraft structures etc.

A loudspeaker is more difficult to monitor in this way because the diaphragm is subject to break-up resonances and phase delays, compared to the motion of the speech coil. This has usually been considered to be the easiest point at which to apply a subsidiary coil or other form of motional pick-up device. Attempts to apply capacitative probes to parts of the diaphragm often failed because of the small voltage obtained and the difficulty of maintaining a stable mean position. Any large pick-up interferes with the motion of the diaphragm to an unacceptable degree.

However, Philips have now made a break-through by fixing a very small piezoelectric pick-up element to a selected part of the cone of a small high-quality loudspeaker, which basically has a diaphragm free from break-up modes. The device generates a voltage due to its own small amount of inertia. In other words, it is an accelerometer which accurately follows the motion of the diaphragm. By feeding the output through electronic integrating circuits, a voltage proportional to the diaphragm velocity or displacement is readily obtained and is applied to the negative feedback loop of the associated power-amplifier.

If one considers the extent of the improvement of the response and distortion characteristics of a transistorized power-amplifier as a result of applying overall negative feedback, it is seen how the performance of a relatively small loudspeaker may be revolutionized. It is potentially possible to obtain a virtually flat frequency response and to maintain a much larger level of bass response without the non-linearity which normally occurs on large diaphragm excursions, due to the non-linear restoring spring force of the diaphragm centring bellows and surround, and the non-linear fringing flux which the voice coil traverses at the outer edges of the magnet gap.

There is obviously a limit to which one can go in securing such improvements, as is seen in the sharp ultimate overload point of a feedback-controlled amplifier, but it would seem that this approach offers a real step forward in obtaining a good performance from a compact moving-coil loudspeaker, which is easily housed in normal domestic surroundings. The concept of a power-amplifier associated with each loudspeaker also conforms with modern multi-channel and quadraphonic stereo trends.

Stereo headphone listening

The availability of a number of high-quality headphones of moving-coil type and several electrostatic designs makes individual personal listening increasingly popular. Certain points may be noted. Firstly, earphones are available in various impedances and it is desirable to take certain precautions when connecting them to audio power-amplifiers. In some cases, amplifiers have headphone outlet sockets provided and the makers advise on suitable types of headphone.

If you are making your own connection to an amplifier, check that the headphones are connected to the speaker outlet and that this is suitably isolated by capacitors etc from DC and the mains, using isolating transformers, if necessary (RS Type USI, etc). Care must also be taken that the level available is not so high as to overload the phones or the listener's ears, even momentarily. Some inserted loss is desirable, so that the headphone level will require a similar volume-control setting to that of the loudspeakers, when these are in use.

Also, many people find that the full stereo effect on phones is disconcerting, in that it places the listener in the centre of the apparent sound source, rather than giving the illusion that it is outside him. A limited amount of cross-mixing between the phones is sometimes thought to produce a more pleasing result. CBS have proposed a frequency-dependent cross-mixing circuit which simulates the head diffraction effect.

A control network has been devised which automatically gives a suitable amount of attenuation for phones in the popular impedance range between 8 and 300 ohms, when fed from normal low-impedance amplifier outputs of 3 to 16 ohms. The network consists of 150 ohms in series with each earphone and 10 ohms in shunt with each. A variable 400-ohm control resistor is connected between each earphone where the 150-ohm resistor joins it. Reducing the value of the variable resistor can give the desired amount of cross-mixing (Fig 61).

Headphone listening has obvious advantages where one or more people wish to enjoy high-quality reproduction without interference from other activities in the same room, and without the disturbance which loudspeaker reproduction at a high level may cause to neighbours. The value of comfortable noise-excluding earpads is obvious. However, there is almost always some loss of low frequencies through a necessarily imperfect seal to the ear, and thus the bass control on the amplifier must usually be turned up, and possibly the high frequencies reduced slightly, in order to give well-balanced reproduction of music, in particular.

The quality of stereo reproduction on headphones of the best types is now so high that one may seriously ask whether there is a

Hi-Fi for the enthusiast

case for concentrating on headphones entirely where only one or two persons normally wish to listen, and to save the considerable expense of high-power amplifiers and loudspeakers of comparable quality. Most headphones may not be efficient enough to operate directly from the output of preamplifiers, but special headphone amplifier designs have been published and are also marketed as complete self-contained units (Shure, Eagle etc).

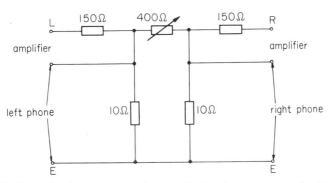

Fig 61. *A simple output control circuit designed to prevent overloading and possible hearing damage when headphones of any impedance are plugged into power-amplifier stereo loudspeaker outputs. The variable control allows for a degree of cross-mixing to be introduced if desired in order to avoid an exaggerated stereo effect.*

The wearing of earphones for long periods is, however, dependent on lightness, low pressure and the softness of the earpads. Some people also object to the somewhat claustrophobic isolation of highly noise-excluding headsets. Membrane radiator types of earphone, such as electrostatic or printed-circuit magnetic ones, have very light-moving systems and require free air access to the rear, and do not usually give a high degree of outside noise exclusion. They may thus meet this objection, in addition to providing very high quality.

Moving-coil stereo headsets

A large number of these are now available at prices ranging from a few pounds to one exceeding that of a smaller cabinet loudspeaker. The units employed vary from small cone loudspeaker units in a rigid enclosure to one or more specialized units designed with domed diaphragms and cross-overs between bass and top units. Obviously, the cheaper types will tend to have a more restricted response, but most moving-coil units are very efficient and will reproduce sound levels as high as one can withstand, at a low distortion level. Note that wire lead colour coding is standardized for stereo phasing, as well as jack connections.

Membrane-type electrostatic and printed-circuit magnetic earphones

Electrostatic phones similar to miniature electrostatic loudspeakers have been available for some while. A constant-charge high-impedance membrane is sandwiched between two perforated electrodes insulated from it. Polarization of about 60 volts is necessary and the electrodes are carefully isolated from the wearer and the amplifier. A separate polarizing and headset control box is normally supplied. The membrane is driven in "push–pull" between the electrodes, with low distortion, and naturally lends itself to the design of high-quality earphones with a soft seal on to the ear and an open back, as it functions substantially as a relatively large well-damped velocity source which covers the front of the ear pinna outer cavity, making it easier to obtain a good match to the ear, together with freedom from resonant effects.

Several manufacturers produce electrostatic headsets (Koss, Stax etc), and also home-construction designs have been published for those who feel qualified to undertake such a project (see *Wireless World*, November 1971).

A very interesting new design is the Wharfedale Isodynamic headset which uses a flexible-printed conductor pattern on a plastic membrane, the whole being sandwiched between suitably-arranged flat ceramic permanent magnets, which maintain a perpendicular flux pattern across the current-carrying conductors. The membrane is thus moved as a whole in a similar manner to the electrostatic system, but without the need for polarization or special feed-circuit connections. The construction is shown in Fig 62. The specification is as follows: *Impedance:* 120 ohms ± 15%; *Power handling:* Maximum input not to exceed 25 V RMS (music or speech); *Sensitivity:* 30 mW input for 95 dB sound pressure level; *Frequency response:* substantially flat 40–17,000 Hz; *Weight:* 450 grammes (16 oz) including cable and plug. Supplied with 3 meters of non-tangle coiled cable, fitted with $\frac{1}{4}$ in. phono-jack plug (wired for stereo).

The membrane is very light and is well damped by the fabric-covered holes in the front and back pole plates, thus giving a flat frequency response and a good transient "attack" or rise time. A relatively high impedance is obtained with the very fine printed conductor tracks on the membrane. This has the advantage that the phones are conveniently operated directly from the loud-speaker output sockets on most amplifiers without much alteration of the volume-control position.

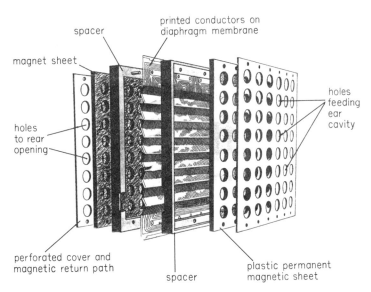

Fig 62. The construction of a dynamic earphone with a plastic-printed conductor track on a ribbed flexible diaphragm, showing an exploded view of the diaphragm and magnet-system assembly (Rank Wharfedale Ltd, Bradford).

Quadraphonic earphone listening

The connection of earphones to four-channel stereo programmes obviously presents a problem, although so-called quadraphonic headsets are marketed. These provide back and front units coupled into each ear cavity, but it seems problematical whether a back and front spatial impression is thereby achieved. It seems likely that a side mixing effect is obtained in the ear cavity. This could be obtained by electrical blending between L_F and L_R, R_F and

R_R in the feed circuits to normal stereo earphones. Nevertheless, there may still be some advantage in the extra power-handling capacity of the four separate units of the quadraphonic headset, and the probability that a better sound distribution is obtained in the ear-coupling cavity due to the use of two diaphragms.

Specialist suppliers of loudspeaker units, kits and cabinet accessories

Messrs Wilmslow Audio, Cheshire SK9 1HF, produce a useful list of many of the best loudspeaker units, kits, cross-over components including coils and reversible electrolytic capacitors, cabinets, fittings, wall brackets, floor stands, etc. They normally stock Fane 13 × 8 in. roll surround units and ITT tweeters for the loudspeaker design shown in Fig 47.

Celestion, Goodmans, KEF, Wharfedale and Radford loudspeaker units and kits are also listed from this supplier, in addition to various horn loudspeakers and designs for housings; complete cabinets are also listed.

"Acoustilux" BAF wadding for damping cabinets and porous neutral-colour foam padding for use behind front speaker cloths, such as the Vynair, are also listed in various colours and patterns.

Messrs Amtron (U.K.) Ltd, Hastings, Sussex, supply a useful range of fluted or square-sculptured acoustically-transparent foam loudspeaker front treatments in various colours, as well as Valcro instant-grip strip for holding fronts to loudspeaker baffles. This type of front is used in a number of high-grade loudspeakers on sale at present. It is probably easier to fit than cloth etc, and is possibly less likely to give rise to unwanted effects, buzzes etc.

Messrs Nichols Acoustical Fitments, Selby, Yorkshire, supply BAF wadding and long-fibre pure wool for acoustical damping. Black foam padding, Vynair, Tygan front cloths supplied in many patterns.

Advanced stereophonic and multi-channel sound-reproducing systems

Almost from the beginning of sound reproduction it has been appreciated that more than one reproducing channel is needed to obtain truly faithful sound reproduction. Systems for use in the home are obviously limited as regards the number of channels which can be used. Hence we are concerned with the various methods now open to us by which practical stereo processing and reproducing equipment may be designed, and arranged in the home to recreate a satisfactory sound field in the listening room. There is often considerable scope for manipulation of the stereo signals in various ways, some of the most important of which will be examined.

Two-channel stereo produces, in effect, sources spread along a frontal horizontal line in the listening room. Present four-channel reproduction using true or "pseudo" quadraphony surrounds the listener with sounds in the horizontal plane. One may ask whether an attempt should be made to pick-up and to reproduce some additional sound in the vertical plane, and experiments have been made to see if it is possible to encode such sounds in a three- or four-channel system.

However, it would seem that most sounds in normal speech and music tend to approach at angles not very far removed from the horizontal, and thus apart from special effects such as aircraft etc, and a loss of some of the vertically-incident sound reflections in the original reverberant ambience, horizontal quadraphony would seem to give a good result within the compass of our present resources. In fact, it is postulated that provided a good "spread" of reverberant sound, both in frequency and in time, is received, the ear is not likely to perceive whether the ambient sounds truly belong to the horizontal or the vertical planes. It is also apparent

192

that there is normally less high-frequency content in reverberant sound, as these frequencies are more readily absorbed in the air and by the natural properties of acoustic-absorbing surfaces in the studio. Rear-channel sound may sometimes be reproduced with a top cut, which has the advantage of reducing any distracting rear reproduction of background hiss.

Thus, in general, the addition of rear-reproducing channels fed with appropriate signals can be expected to represent a significant advance on two-channel frontal stereo in certain important respects. However, it is possible that the addition of the extra channels for quadraphonic reproduction of sound from four or more loudspeakers distributed around the listener, dilutes the sound-image location, particularly if omnidirectional loudspeakers are used together with off-centre listening positions.

It may be fortunate that a great deal of our most loved traditional music, particularly by composers such as Bach, Verdi, and other masters, was written for performance by large groups of musicians, singers etc, who were necessarily quite widely distributed in the original church or hall. The audience were intended to have a sense of being involved in and surrounded by the whole performance. Quadraphony or "surround sound" is likely to help this type of illusion considerably. Put at the minimum, it might be regarded as offering a "trade-off," in which some reduction in positional sharpness is accepted in return for a more satisfactory enveloping ambience and an enhanced sense of involvement.[24]

Stereo circuit developments

The basic aim of all stereo systems is to recreate in the listening room a closer approach to the original sound field than is possible with single-channel monophonic systems. Some spatial limitations due to the difference in scale between a concert hall or studio and a living room must always remain. However, stereo equipment and circuitry are continuously being improved; in particular, quite radical improvements have been made by the availability at relatively low prices of integrated circuit packages which perform the often very complicated circuit functions involved in stereo decoding and signal processing. As recording media are improved and the scope of micro-electronics grows, we can look forward to further developments in the future.

We have progressed from two-channel to multi-channel stereo

in that discrete or multiplexed quadraphonic broadcasting and recording systems are becoming available, as well as ingenious matrixing systems which carry four-channel information on the currently available two-channel media. It is also possible to generate "pseudo-quadraphony" or "surround sound" for one-self by using circuits operating on any normal two-channel stereo programme. We may now consider some of the principles, circuits and equipment currently available for general use.

Stereo system designation

The way in which stereo signals are encoded and carried by the various media and systems is now so complex that a form of shorthand has been evolved to show the channel sequence in the various systems. Thus 1–1–1 denotes a monophonic system in which one final input (which may be mixed from several micro-phones) is carried by a single-channel medium such as a mono-phonic record groove, and is finally reproduced by a single amplifier channel, which may of course still be fed to several loudspeakers in parallel.

A discrete or separate two-channel stereo system is then a 2–2–2 system, because two stereo input signals are carried by two separate channels, which may still be adjacent in the medium. Examples are two tracks on a tape or on the two walls of a stereo record groove. Although the channels are theoretically separate, some mutual cross-talk or mutual interference may occur to reduce the stereo separation. The two tracks are normally reproduced on two separate left and right loudspeakers or groups of loudspeakers.

A discrete four-channel quadraphonic system such as a tape-recording is a 4–4–4 system. A "pseudo-quadraphonic" surround sound system where four channels are derived from a two-channel stereo programme is a 2–2–4 system, while a matrixed system, where the original four-channel quadraphonic signals are carried by a two-channel system such as a matrixed gramophone record, is described as a 4–2–4 system, as the four channels are reconstituted by the decoder matrix.

Multiplexing and matrixing concepts

We are familiar with the idea of making one transmission medium carry more than one separate programme or information channel.

One example is the ether, which is filled with myriad radio radiations which can be tuned in because they are separated into individual discrete frequency bands. Any main transmission channel can be made to carry discrete subsidiary channels if these are modulated on to separate carrier frequencies and if there is sufficient frequency band-width available in the channel to pass all the separate frequencies comprising the full modulation concerned.

The general technique is known as "multiplexing" and the degree of "discreteness" or separation of the various channels transmitted simultaneously depends largely on the frequency spacing and the degree of selectivity of the tuning circuits used, as well as the precise type of modulation adopted.

These methods are often expensive in terms of equipment and circuitry and the actual frequency band-width which has to be used.

Economies in the required band-width may be made by suppressing certain sidebands, or suppressing carrier frequencies in transmission and restoring these later at the receiver. FM stereo broadcasting is economically carried out by modulating the sum of two stereo channels on to the main FM "baseband," while the difference signal is modulated on to a suppressed sub-carrier, which is reconstituted in the receiver decoder circuits by a trans-mitted pilot tone, in the manner outlined in Chapter 2. Alternative-ly, a relatively simple method of combining signals and later separating them at will is that of "matrixing." An electrical matrix is an array or grid of cross-connecting elements between various input and output circuits, whereby the incoming and outgoing signals are mixed together or separated according to the laws governing the circuit elements used (typically resistors, capacitors and transistors, connected in a regular fashion between the various lines). The action of matrixing and de-matrixing circuits may be analysed and described by simultaneous equations and matrix algebra which, although applicable to electrical matrices, is a separate subject. This may cause some confusion to those new to the studies in question.

Matrixing in a particular form is used to obtain the stereo sum and difference signals subsequently processed as discrete multi-plexed signals in FM stereo broadcasting and reception. Matrixing is basically a relatively simple process, and in the case of two-channel signals and idealized multi-channel matrices, complete ultimate separation of the matrixed signals may in theory be achieved. In practice, as one would suspect, there are output errors and "cross-talk" between the signals. Some errors are due to component tolerances and imperfect balance of the circuits; others may be due to the basic limitations of the circuits one is able to use in practice. The use of complex integrated circuits is producing improvements. Fig 13 shows the multiplexing spectrum used in stereo FM broadcasting, broadly similar techniques being usable for two and four channels.

For the de-multiplexing filter circuits and the de-matrixing circuits which follow the discriminator in a typical stereo broadcast receiver decoder section, a combination of both these techniques is used in order to achieve the final separation of the two stereo channels. Due to the residual errors mentioned, some cross-talk between the L and R channels is introduced. This is at a low enough level to be neglected in a good quality decoder. SQ and QS matrixing are also sometimes used now for quadraphonic broadcasting.

Four-channel quadraphonic broadcasting is also carried out in the USA by using the Dorren principle. This is a development of the present two-channel FM system described, but uses additional matrixing and an extra 90° quadrature multiplex sub-carrier, and a 76 kHz sub-carrier to carry the difference signals which arise. Similar filters and matrices in the receiver decoder produce the four stereo quadraphonic channels. It is claimed that little extra FM band-width is required, and that the reduction in signal-to-noise ratio is not much more than is normally experienced with two-channel FM stereo broadcasts.

Four-channel quadraphonic gramophone discs are now recorded on several different matrix systems and on at least one multiplex system. Some of the various systems in widespread commercial use are described in the following pages.

Discrete four-channel systems

Full quadraphonic discrete programmes are available from quadraphonic tape-cartridges or four-channel cassettes produced by JVC, RCA, EMI, Decca, and other companies. With the rapid growth in tape-cassette recorder/reproducers of a quality approaching that of the best discs and reel-to-reel tape-recorders, it is certain that more and more pre-recorded cassette programmes will be sold, and that increasingly more amateurs will record their own original performances. There is much scope for experiment with four-channel microphone arrays to discover if good quadraphony can be achieved. Four cardioid unidirectional microphones mutually angled at 90° to give a "clover-leaf" pick-up pattern are often used. Some degree of adjustment of the angle between the frontal pair of microphones may be desirable in order to control the apparent width of the frontal sound source.

Four-channel reel-to-reel discrete recorders, with four separate record/replay amplifiers, are often favoured by amateur and professional recordists because of the relative ease of tape editing, as compared to cartridges or cassettes. It is possible, with some experience, to edit cassettes, and some amateurs may prefer to use cassette machines exclusively because of the availability of pre-recorded programmes on cassettes as opposed to reels. Alternatively, it is possible to re-record the final edited tape on to a cassette or cartridge, assuming that the appropriate machines are available.

Four-channel systems and matrixing principles

In a basic four-channel matrixing and de-matrixing system, the four discrete original stereo input signals are fed into a network with four symmetrical arms, comprising resistors and transistors set to give transmission gains or losses, which are proportional to the figures which refer to each arm of the matrix. The two output channels resulting from the matrix may then be transmitted or recorded on the two discrete channels of a stereo gramophone disc etc. The signals are mixed without any increase in band-width and the disc may thus be replayed by a normal stereo gramophone pick-up. The two output channels from the pick-up may be fed into a similar de-matrixing circuit. The outlets of this complementary matrix produce four stereo output channels which are applied to the L and R front and rear loudspeakers in the same way as in any other form of quadraphonic reproducing layout.

However, a little observation shows that in a simple matrix system such as the above, if fixed amplitude changes only are imposed on the input signals, a signal on either one of the incoming two channels will produce some output on all four outputs. In other words, inter-channel cross-talk has been introduced. The reason for the adoption of particular fractional values for the attenuation in the matrix arms is to maintain an equal total power distribution between input and output circuits as a whole.

It was soon realized that various artifices could be adopted in order to reduce or substantially eliminate the unwanted cross-talk between channels. The basic properties of audio signals, such as instantaneous amplitude variations, frequency spectrum, rise time of transients and the phase relationship between the component parts of the waveform, must not be altered by matrixing, but both the amplitude ratio between the channel signals and the phase displacement relative to each other may be deliberately changed in the matrix and then used in the de-matrixing process to sort out the four channels again with the minimum of cross-talk, provided that the relative amplitude and phase displacements are then corrected to bring the signals back into their original relationship.

The auditory precedence effect can also be exploited by incorporating circuits which introduce a phase reversal or a phase delay on the unwanted cross-talk signals appearing in a channel due to matrixing errors, thus rendering them relatively harmless as regards their effect on true image location, but enabling them still to add to the general background ambience. Ingenious quadraphonic stereo "enhancement logic" circuits have also been devised which monitor the channel signals continuously and instantaneously change the matrixes and/or the associated circuitry so as to produce, in the final quadraphonic outputs, the nearest possible similarity to the original discrete four-channel stereo programme. This type of approach is used in the CBS SQ matrix system, now widely used by EMI and other gramophone record companies. Different circuits are used in the Sansui QS Variomatrix system on some other records. These latest methods are widely acclaimed and give improved separation without the need for any increase in band-width or the use of non-standard pick-ups. Additionally, the records are compatible with mono or two-channel frontal stereo reproduction, in that one- or two-channel reproducers can give a satisfactory result.

It is thought that the major recording companies are now recording all important master tapes as four-channel discrete programmes and it seems likely that, eventually, all gramophone records will be issued as quadraphonic discs. Discrete quadraphonic tape-machines and cartridges are now also sold.

However, it is well known that quadraphonic gramophone-record matrix systems are not standardized, in that some companies, for instance, produce records on the QS and the RM or regular matrix systems. One must distinguish the JVC Nivico

CD-4 multiplexed discs which use a modulated supersonic sub-carrier and require a special pick-up and stylus, to give what is said to be discrete four-channel performance.

The public may be justifiably confused as to what type of records and which equipment to use. Fortunately, it seems that there is a reasonable measure of compatibility between many of the systems in that, even without the optimum type of decoder for a given system, one may enjoy some of the benefits of quadraphony, particularly as regards improved ambience. Also, integrated circuit "building blocks" come to the rescue, in that these have been specially designed and are available for most systems.

The first four-channel systems on the market were arranged to operate from any two-channel stereo programmes and offered "pseudo-quadraphony" or "surround sound" by using synthesizing matrices which produce two phase-displaced rear channels derived from the front channels of normal two-channel frontal stereo. A pleasant and to some degree controllable increase in ambience is obtained which is found to enhance the enjoyment of many programmes.

Fig 63 shows details of an integrated circuit matrix system which is designed to give this type of ambiophonic reproduction.

A useful approach to adopt in present circumstances is to use separate printed-circuit modules built around the various ICs for the particular matrixing systems used. This has been done by Quadraphonics Ltd, who provide interchangeable matrix-circuit modules on edge-connector boards which can be plugged in as required by the users of their amplifier system. Other manufacturers are supposedly adopting a similar approach to the problem.

Thus the investment in four power-amplifiers with suitable volume, balance and tone controls in the associated preamplifiers, together with four good-quality loudspeakers, would seem to be a sound move for an enthusiast to make.

It is worth noting that, although some reduction in the power required from each amplifier may be possible, it is a mistake to down-grade the rear amplifiers and loudspeakers, as any distortion or coloration introduced will probably detract noticeably from the overall fidelity. Some additional top-frequency cut or roll-off may, however, be applied to rear channels to remove any residual noise. Noise may be more distracting from the rear and, in any case, it is likely that normal room reverberation has reduced high-frequency content due to acoustic diffusion and absorption of high

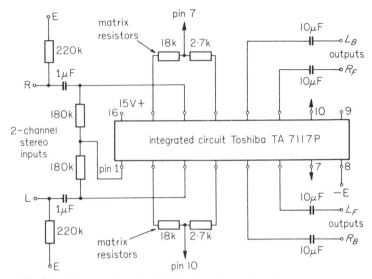

Fig 63. *A circuit using the Toshiba four-channel synthesizer integrated circuit TA 7117 to produce left and right front and rear signals for "surround sound" feeds. The inputs can be derived from any two-channel stereo input signals (from tuner, disc etc). The IC includes differential amplifiers on the L and R input lines, which feed 0° and 180° inverted signals to a matrix circuit whose coefficients can be adjusted by the values of the external resistors. The four outputs from the matrix may be fed direct to four power-amplifiers and loudspeakers placed around the room as for quadraphony (supplied by* Toshiba UK *via* Erie Electronics Ltd, Yarmouth).

frequencies, so that a slight additional top loss introduced on lower-level high frequencies will not be perceived.

The CBS SQ quadraphonic matrix system

As described above, this system seems to offer many advantages such as image separation and a good general quadraphonic effect, together with the great advantage of compatibility with existing equipment of monophonic or two-channel stereophonic types. Integrated circuits are made for SQ decoding and designs have been published giving constructional details and circuit values for the extra passive components required on the associated printed-circuit mounting boards. A number of suppliers also provide complete kits or assembled boards. Details are given in this section of circuits for ICs. The constructor can thus add the appropriate controls, with four amplifiers and loudspeakers, so as to provide a complete matrix SQ quadraphonic reproducing system, based on the ICs.

Fig 64 shows a basic SQ module built around an IC made by Motorola. The present application is mainly to CBS, EMI and

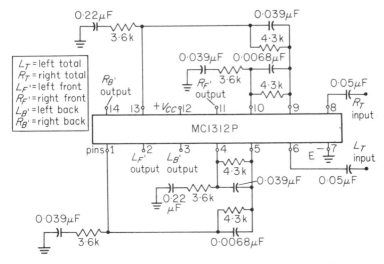

Note:— for optimum performance ±5% tolerance components are recommended with the exception of the input capacitors

Fig 64. Basic SQ quadraphonic decoder using the Motorola MC 1312 IC. Blend resistors of 47 k-ohms and 7·5 k-ohms respectively may be added across R_F, L_F and R_B, L_B. Alternatively, additional ICs MC 1314 and 1315 may be added to give logic-enhanced separation between channels (SQ is the trade mark of CBS Inc.) (Motorola Semiconductors Ltd).

other quadraphonic SQ matrixed gramophone records, but SQ encoded cassettes, cartridges and FM broadcasts are being tried in some countries, as convenient means of making quadraphonic programmes available on two-channel media. Stereo-enhancement logic circuits are also available as ICs which can be added to the basic SQ decoder in order to improve the matrix action in regard to image separation and placement.[27]

The QS and RM matrix quadraphonic systems

Sansui and other manufacturers have adopted different matrix arrangements which require decoders designed for the specific systems concerned if optimum results are to be obtained from the appropriate records. Records on the QS system include Pye issues.

Integrated circuits and modules for making QS decoders are obtainable from Japanese and US firms. Some of these also include "vario-matrix" stereo-enhancement circuits in order to give improved separation and image location.

The CD-4 multiplex gramophone-disc system

The JVC Company of Japan and RCA have developed this multiplex system of gramophone-disc recording and reproduction which is an alternative to matrixing systems for quadraphonic reproduction. It uses a modulated sub-carrier at a frequency of 30 kHz, carried on each groove wall, in addition to an audio frequency baseband, in a manner rather similar to stereo multiplex broadcasting. Fig 65 shows the way in which the sum channels are

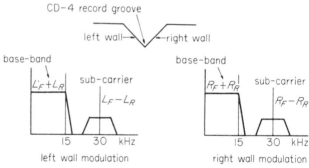

Fig 65. *The JVC Nivico CD-4 quadraphonic four-channel multiplexed gramophone-disc system shows how the two front and two rear signals are modulated on to the two walls of the groove, the sum signals being carried in a baseband and the difference signals by the modulated 30 kHz sub-carrier.*

carried in the baseband and the difference channels by the frequency-modulated sub-carrier. This is a sound communications technique, but it involves an advanced form of gramophone recording and pick-up reproducing technique. The pick-up output is amplified in a wide-band audio and supersonic amplifier and then passed through a demodulator to special matrix circuits which separate the signals into the four output channel amplifiers for application to four loudspeakers. Disc recording and reproduction up to 50 kHz requires the following special measures

1 Due to cutter head and cutting stylus limitations, the discs are cut at half the normal speed of $33\frac{1}{3}$ rpm, speeded up for re-play. This means downwards frequency translation on recording, requiring extra-low frequencies to be cut for a normal bass response.
2 Great difficulties obviously had to be overcome in devising a pick-up stylus and mechanical system which will track a standard record groove which is modulated up to 50 kHz. (The normal pick-up and disc-system response is maintained only up to 20 kHz, which is, of course, quite sufficient for normal audibility.) Key questions were resonances, including resonance between stylus mass and disc-material stiffness, high-frequency tracking difficulties due to a finite stylus mass, and wear of the high-frequency groove modulations. It is necessary to use specially-shaped miniature ovoid styli (eg that attributed to Shibata). The high-frequency modulated band is recorded nearly 20 dB lower than the normally recorded audio frequency baseband, and specially toughened record-pressing material is used.
3 Compatibility with mono or normal two-channel pick-ups and systems is a commercial necessity for any record system. This is obtained automatically, in that the audio baseband on the two groove walls carries the stereo sum signals $L_F + L_B$ and $R_F + R_B$, which will usually combine or "fuse" audibly to give a satisfactory result when the special records are reproduced with normal pick-ups. Some wear and increase in noise of the upper band is to be expected if a heavier (say 3 g) pick-up is used instead of the 0·5 g pick-up specified for the CD-4 discs.

The demodulation process is critical and uses integrated circuits operating on the phase-locked loop principle. Demodulators are produced as separate self-contained cabinet units for use with existing amplifiers, as well as part of complete four-channel amplifier systems (JVC (UK) and agents distribute JVC Nivico CD-4 equipment). European manufacturers (eg Cosmocord) are stated to be producing high-frequency pick-ups suitable for CD-4 records, which are available with an increasing number of titles from JVC, RCA and Warner, etc.

The CD-4 system is basically capable of giving quadraphonic sound separation which approaches that of the original four-channel programmes in the form of discrete tracks on tape from which most discs are made. It obviously represents a carefully engineered system of great precision which, like other Hi-Fi equipment of the highest quality, has to be treated and maintained with appropriate care.

The UD-4 multiplex 4-channel disc system

This is a later system attributed to Nippon Columbia (Japan) and D. H. Cooper of the University of Illinois. It is basically described as a carrier system like CD-4 which carries baseband audio signals on each groove wall as well as a modulated carrier of a supersonic frequency. Instead of using the left and right sum signals in the basebands, the UD-4 proposals use SQ or QS matrixed 4-channel signals L_T and R_T in the left and right hand walls of the record groove, respectively. Two further sum and difference signals are modulated on to the 30 kHz carrier on each groove wall. The basebands occupy a wider spectrum (up to 20 kHz) and the carrier sidebands a narrower spectrum than CD-4, the result being an overall economy in band-width of about 10 per cent, slightly easing the high-frequency pick-up problems, it is claimed. If the records are played (*a*) into a matrix decoder called the BMX (baseband matrix), rotationally symmetrical 4-channel signals result which have a reasonable measure of separation. (Rotational symmetry means that the directional sound pattern produced by the four decoded channels in the listening room remains constant if the original sound source is panned completely round the microphone pick-up point in the recording studio.) If (*b*) a UD-4 multiplex demodulator is used, rotational symmetry is retained, but enhanced separation is achieved between the four output channels to give an approach to discrete channel performance, as with CD-4. Good monophonic and two-channel stereo compatibility are claimed, as well as the options of using a simple matrix decoder (*a*) or a more complex demodulator (*b*), as above.

Another new stereo encoding system is the British NRDC Ambiosonic proposal, which encodes vertical as well as horizontal sound components picked up by a special microphone array.

At present, recordings made by the above two new systems are not apparently commercially available. It remains to be seen in the future how far the various systems we have described can be brought together, instead of being used more or less concurrently, as at present.

Planning a complete high-quality system

If a complete installation is being planned from scratch it is useful to have a guide as to the best way to apportion one's funds so as to obtain a system of balanced quality.

As a very rough indication, it might be advisable to budget so as to spend approximately equal sums on—

1 The loudspeaker(s) and enclosure(s).
2 The amplifier system, including preamplifiers.
3 Gramophone turntable and pick-up.
4 Tape-recorder (if required), reel-to-reel, cassette or cartridge.
5 Radio-tuner unit and aerial system.

It is, of course, possible to spend more or less money, according to the quality, power and facilities desired.

If the enthusiast is prepared to construct his own loudspeaker enclosures and other cabinet work a considerable saving will be made. If he is able to construct his own amplifiers, tuners and other circuitry, a proportionately larger saving in cost may be achieved. Other compromises between quality and cost can be worked out according to the power output required from the amplifiers and loudspeakers and whether, for example, extra features such as a high-grade automatic record-changer could be dispensed with in favour of a single-record player or whether a tape-recorder can be dispensed with.

In the following pages we illustrate some suggestions for accommodating possible systems designed to fit small and large rooms with various degrees of complexity and variety in the quality and number of services offered.

The housing of Hi-Fi equipment and units involves problems concerned with acoustics, power supplies, ventilation, ergonomics (the science of human control of machinery), aesthetics (appearance and form), space, portability (weight) and safety. This is a formidable list and in modern housing and apartmental living conditions it is obviously extremely difficult to find space for a technically well-planned system. Given enough space, one can obviously design large pieces of furniture or built-in cabinets which can house any desired amplifiers, tuners, turntables, tape-recorders, loudspeakers etc. Many examples of luxurious installations have been described in the Hi-Fi journals. They are often the work of dedicated enthusiasts who have been able to devote a fair amount of money and a large room to their hobby.

The average Hi-Fi enthusiast is not in this happy position and almost invariably has to make a number of compromises.

The possible approaches vary considerably, but they may generally be classed either as designs in which most or all of the equipment, usually with the exception of the loudspeakers, is housed in one enclosed cabinet or, alternatively, the various units may be housed separately. It is difficult to integrate such an assemblage in a completely satisfactory manner owing to the differing sizes and shapes of the various units and the need for access to the controls and mechanisms. It is natural that the "bookshelf" type of approach should be gaining favour, as it offers a number of advantages to the amateur.

Equipment housed in cabinets

Fundamental decisions on the designs relate to the question of how the user intends to operate the equipment, and whether his approach is that of the armchair listener who wishes for the minimum disturbance of his comfort or, alternatively, if he regards himself more as a technical operator who is prepared to cross the room and operate controls etc from a standing position. A number of constructors' designs and fully assembled cabinets have been produced which give a choice of arrangements of both kinds. In considering the merits of any particular design, such factors as the position, angle and lighting of controls and mechanisms for ease of operation from a sitting or standing position must be considered, together with considerations of bulk, appearance and the enclosing of equipment with doors and covers in order to keep it free from dust etc when not in use. It is generally well worth using two or more separate loudspeaker cabinets. The loudspeaker enclosure can then be correctly designed in its own right, the risk of microphonic feedback to pick-ups is largely eliminated and correct loudspeaker placement in the room is greatly facilitated, particularly for stereo. Figs 66–70 show some interesting commercially-produced arrangements offering space for tuners, amplifiers, turntable, and in some cases tape-recorder and record-storage etc. Efficient dust protection is generally afforded when the units are not in use.

Fig 66. A recent compact high-quality stereo tuner and amplifier plinth assembly which includes the BSR high-fidelity automatic record-changing turntable normally fitted with a smoked Perspex (tinted) lid. Two special omnidirectional loudspeakers shown here are wide-range column types which handle the output of 15 watts per channel.

Fig 67. An amplifier and turntable plinth assembly using the high-quality Garrard AP 75 automatic single-record-playing turntable and arm assembly on a plinth into which Sinclair stereo preamplifier and power-amplifier modules have been assembled. The modules of Figs 34–9 may be used or alternative power units may be fitted to give power output ratings of over 40 watts per channel. The AP 75 arm gives a micro-adjustment of tracking weight down to below 2 g with side thrust compensation. Automatic record-starting, groove-positioning and automatic lowering, raising and stop are provided as well as a pause-control.

Wall and room-divider types of housing

These have considerable appeal today, as many tuners, amplifiers, tape-machines and loudspeakers are available as free-standing units in their own cabinets. The size of all these units is now such that they can usually be accommodated in the "bookshelf" or "divider" type of structure. A number of commercial kit systems are available for making up those units to individual require-ments. The vertical open-work structure is economical of room

Fig 68. *A convenient trolley type of cabinet (Imflex Mini-trolley) for tuner, record-player and amplifiers, which includes some storage space. A hinged Perspex lid covers the units when not in use. The height is convenient for armchair or normal operation.*

space as it keeps the floor clear. The systems are versatile, in that alterations and extensions to the system can generally be accommodated and the general effort can be made to harmonize with modern room designs and decorative schemes. It is desirable, however, to run the various interconnecting wires and cables neatly and preferably as far as possible out of sight behind structural members etc. Fig 70 shows wall and divider units of the kind described above. Shelving by Click Ltd is also sold.

Fig 69. *A compact and reasonably priced cabinet (the "Trio") for housing tape-deck, turntable, tuner and stereo amplifiers as well as providing alternative storage space.*

Detailed system planning

Having decided on the general scope and the approximate sum of money available for the system, one must consider precisely how it is to be realized. For example, a tuner and amplifier stereo system and gramophone turntable system with two good loudspeakers, all in the form of separate but technically compatible units, may be decided on, with the possibility of operating later with a tape-recorder and microphone. Use with slide or cine-film sound accompaniment may also be considered as a later project.

The obviously safe ways of buying recommended equipment from reliable dealers have been mentioned. It is of great importance to ensure that a major investment of this kind is not only going to be technically compatible as an assemblage of units but will

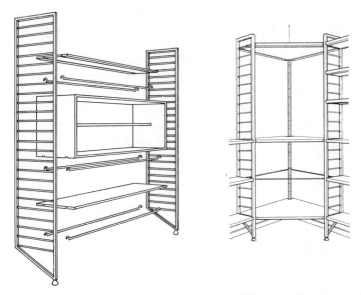

Fig 70. *A comprehensive wall-fixed or room-divider unit (the Staples "Ladderax" unit) giving a flexible arrangement of modular shelves, panels and cabinets to accommodate a wide range of Hi-Fi and other home entertainment equipment, together with books, ornaments, tape storage cases, records etc.*

also really suit your needs and tastes, to say nothing of the space available for housing and laying out the system.

Obtaining the very best equipment throughout is necessarily expensive and the cost of a complete outfit as above could now be £800 or more. On the other hand, obviously the cheapest commercial Hi-Fi equipment must have certain limitations.

Many enthusiasts will wish to save money and also to have good equipment by doing as much as possible of the construction themselves. The obvious possibilities of loudspeaker and other cabinet-making have been mentioned. The use of cabinet kits and ready-to-assemble units is an alternative to the rigours of the carpentry and cabinet-making from scratch which are otherwise involved. A number of such kits are now advertised by reliable makers.

It is worth while to consider building your tuners and amplifiers from established designs, kits or modules, a few of which we have illustrated in previous pages. These are now available from a number of reputable suppliers and we will now consider some of the factors involved in linking up and using units of this kind.

The integration of kits and modules

Apart from the housing of complete and finished amplifiers and other units in bought or home-made cabinets, the use of finished modules or sub-assemblies, the building of kits supplied and packaged to make up approved designs and, finally, the building of amplifiers or tuners to designs published in the technical journals all represent possible steps in the home construction of Hi-Fi equipment.

It is necessary to remember, however, that it is required to assemble, wire and interconnect finished commercial amplifiers and tuners or modules in accordance with the makers' recommendations, so as to ensure that the correct results are obtained. This particularly refers to freedom from hum, interference, unwanted break-through and cross-talk. The correct types of screened connecting leads must be used for aerials and audio-input leads, cables etc in order to avoid excessive capacity losses and so on. Professional advice from the supplier or dealer will enable the correct procedure to be followed. General rules are always to use cables of adequate size and diameter; proper shielded co-ax or twin low-loss RF cable of the correct impedance being used for VHF aerial leads, without any joints or dead-end extensions. These may cause disastrous internal reflections or standing waves on the cable, unless the proper fittings, plugs and terminating devices or baluns are used.

Audio-input connections must be as short as possible and must be kept away from output leads or mains cables. Low-impedance input connections may be run in shielded-twin-microphone cable or thinner equivalents, but high-impedance inputs such as those for the standard 47 k-ohms pick-up inputs should be run in low-capacity shielded co-ax or other cables, unless the runs are so short as to avoid any shunt capacity losses of the higher frequencies. Earthing of cable shields, plug shrouds and the chassis via the mains third pin is of paramount importance. The correct choice of earthing points and the avoidance of hum-causing loops are among the most difficult decisions one has to make in the wiring of audio systems. To some extent, every system is different and it is difficult to give any firm rules. Generally speaking, good low-resistance earth connections must be made everywhere, the main earthing point being applied at one place only; each shield end or other earth connection is run back to this point, to avoid common loops of earth-lead in which small mains leakage capacity currents etc can couple to high-gain input points and cause hum. It is also important to ensure that the metalwork of cases, panels and turntables is earthed properly, not only to prevent hum, but as a safety measure.

Suppliers of amplifiers and complete kits usually include the necessary plugs and connectors with the exception of mains plugs, together with some wiring instructions, but the makers of subsidiary units such as turntables and tape-recorders often do not supply connector plugs, except possibly with a microphone etc. Coaxial "gram plugs" are standard and may be obtained colour-coded, but a number of DIN plugs and connectors are also used and it is important to obtain exactly the correct pin arrangement. One final point is that loudspeakers must be wired with suitably low-resistance leads so as to avoid a loss of power.

Any difficulties one may have in laying out and connecting up finished commercial amplifiers are naturally likely to be less than those which may be encountered when assembling modules or kits of one's own choosing. Here again, reputable suppliers are at pains to give very detailed hints and instructions for various types of set-up. but, of course, the more one follows different ideas the greater the risk of meeting difficulties. However, the makers of good kits and modules are usually very helpful in cases of difficulty, and are generally prepared to service or to correct faults and defects encountered by genuine home constructors, at a nominal charge.

The choice must be yours. Established commercial units will give you a known and usually high standard of performance. Modules and kits will normally be cheaper, and in many cases can claim a very good performance. The precise extent to which this is achieved in practice may depend largely on individual skill. There is also the often very great satisfaction of making and assembling as much as possible of one's own Hi-Fi system, which can then reflect personal tastes and ambitions to a considerable extent.

Names of suppliers of established kits and modules appear in the advertisement and other pages of the Hi-Fi journals and their products are often reviewed, in addition to the normal fully assembled systems.

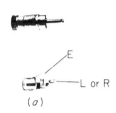

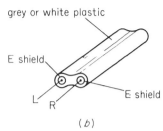

grey or white plastic

E shield

L or R

E shield

L R

(a)

screened phono plugs and socket

(b)

twin stereo screened audio cable

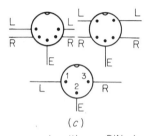

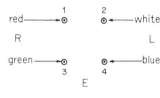

L

R

L

R R

L

R

E

E

1 3

2

L   R

E

(c)

screened multi-way DIN plugs
and sockets (showing connections)

1   2
red————•⊙   ⊙•————white

R   L

green————•⊙   ⊙•————blue
         3   4

E

view of cartridge pins
(d)

stereo pick up wire coding

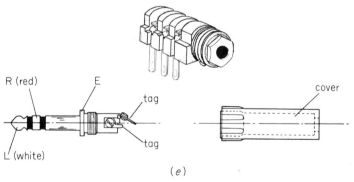

R (red)   E

tag

cover

L (white)

tag

(e)

stereo headphone plugs and sockets with wiring coding

Fig 71. *Commonly used audio connectors and cables.* (*a*), (*b*), (*c*), **RS**
Components; (*e*) Rendar Ltd.

Glossary of Hi-Fi and radio terms

Acoustic Feedback. An undesirable oscillation or howl caused by sound energy from a loudspeaker either being picked up by a microphone or causing a turntable and pick-up to vibrate.

AF (Audio-frequency). Frequencies within the range of human hearing. Generally defined as 20 Hz to 20,000 Hz. (See "Hertz.")

AFC (Automatic Frequency Control). A circuit designed as a refinement to a radio receiver to keep it accurately tuned to the desired signal.

AGC (Automatic Gain Control) or AVC (Automatic Volume Control). A circuit which automatically keeps the audio output of a radio tuner at the same level whatever the strength of the radio signal received.

Ambience. The persistence of sound in an auditorium which gives it a particular acoustical character. (See "Reverberant Sound.")

AM (Amplitude Modulation). The usual method of carrying an audio signal on a radio-frequency wave for transmission. It is the method used in most long-, medium- and short-wave broadcasts. The RF carrier amplitude is varied.

Ambient Noise. Surrounding noises, other than the sounds one is trying to reproduce, ie traffic noise, factory machinery, rain, wind etc, which may affect the microphone or may be present in the listening room.

Attenuation. The reduction in the level, loudness or strength of a radio or audio signal due to deliberate or accidental losses.

Attenuator. A device for providing attenuation (see above) of a signal. The simple attenuator is formed by two resistors. A volume-control is a simple form of variable attenuator. (See "Potentiometer.")

Audio. Relating to sound reproduction and recording.

Baffle. A large flat board on which a loudspeaker is mounted. It has to be large to allow good reproduction of low frequencies. Enclosed cabinets generally give satisfactory performance at a much smaller size. (See "Infinite Baffle.")

Balun. This is a device for offsetting the effects of unbalance when interconnecting UHF or VHF aerial cables. A half or quarter wavelength of line may be formed into a spiral and housed in a box for this purpose.

Bass. The lower notes or frequencies in the audio range.

Bias. In tape-recording, bias is the high-frequency signal (100 kHz approx) applied to the magnetic tape together with the signal to be recorded. This bias ensures that a strong, low-distortion audio signal is recorded on the tape. In valve or transistor circuits it is the small DC voltage applied to the grid, cathode, emitter or base etc to ensure that low-distortion, efficient amplification is achieved.

Bias Compensation of pick-ups is the provision of a small torque to offset the tendency for the stylus to experience an inward force due to the record drag and an offset angle in the pick-up arm.

Binaural listening involves the use of both ears, either directly in a free sound field or by the use of earphones.

Bulk Erasure. A method of erasing or removing the magnetic signal from a complete reel of tape in one quick operation. The reel of tape is passed through the AC magnetic field from a mains-energized electromagnet.

Capacitor (Condenser). A component made up of conducting plates separated by insulation. It presents an increasing restriction to the flow of alternating current as the frequency of the current gets lower. It will not pass direct current. The names given to the various types (paper, mica, ceramic, electrolytic etc) generally refer to the type of insulation material used in the construction.

Capstan. The motor-driven spindle or pulley in a tape-recorder which drives the tape at the required speed. It has to be very accurately machined to ensure the correct tape speed with the minimum of wow or flutter. (See "Pinch Wheel.")

Carrier. A radio-frequency wave which is modulated so as to carry audio-frequency or other signals. (See "AM" and "FM.")

Cartridge. The part of a pick-up which translates the mechanical vibrations of the stylus in a record groove into an electrical signal to be fed to an amplifier. It is fitted in the head on the pick-up arm. Tape cartridge.

Cassette. A miniature container of magnetic-tape-recording spools which is conveniently located on to a standard cassette machine. Magnetic-tape-cartridges use an endless loop of tape and are used on different machines.

Coaxial cable. A screened sheathed cable enclosing a single insulated conductor. A characteristic impedance of 75 ohms is used for VHF aerial and input cable.

Compliance. The reciprocal of stiffness. It is a measure of the flexibility of movement of a pick-up stylus or loudspeaker cone etc. The higher the compliance the more free or flexible is the movement. High compliance is desirable in a lightweight Hi-Fi pick-up or a low-frequency loudspeaker cone suspension. (See "Stiffness.")

Condenser. See "Capacitor."

Cross-over Network. A circuit made up of capacitors and inductors which divides the audio output of an amplifier into two or more bands of frequency, so that all lower frequencies can be fed to a loudspeaker specially suitable for reproducing low frequencies and the higher frequencies to loudspeakers specially suitable for reproducing higher frequencies.

Cross-over Frequency. The frequency at which a particular cross-over network separates the lower-frequency sounds from the higher ones.

Cross-talk. A break-through by part of an unwanted sound signal on to the wanted sound. For example, some left-hand stereo sound is heard on the right-hand loudspeaker together with the proper right-hand sound.

Cycles per Second (c/s or cps). The unit of measurement of frequency rate of vibration. The rate of vibration of a sound wave determines the pitch. Now generally replaced by the hertz, kilohertz, megahertz etc (abbreviated Hz, kHz, MHz etc).

Decibel (dB). A unit of measurement using the logarithmic ratio of two signals. Most commonly applied to electrical signals, sound levels, gains of amplifiers, loss in attenuators etc. dB are added or subtracted where gains are multiplied.

Decoder. In stereo, the circuits which separate the multiplexed or matrixed signals into the required left and right channels for output to stereo amplifiers.

Detector. The circuit in a radio receiver which converts the modulated radio-frequency signal into an audio-frequency signal for application to an audio-amplifier.

Diaphragm. The sound radiating or collecting surface connected with the transducer or energy-converting element of a loudspeaker or microphone.

DIN specifications. German Industry Standards which have been widely adopted for various Hi-Fi plugs and connectors. The DIN 45500 specification also covers many aspects of Hi-Fi equipment and performance. Corresponding BS and international specifications are also being introduced.

Dipole. A symmetrical dual-rod aerial for VHF etc reception.

Discriminator. The RF demodulator or detector stage in an FM receiver.

Distortion. If the signal out of an amplifier differs in any way (other than in level) from that being put into it, distortion has occurred. The commonest form is harmonic distortion or intermodulation distortion.

Dolby System. An automatic method of reducing recording noise and hiss by controlled volume compression and expansion in appropriate frequency bands during the recording and playback process, particularly of master tapes used to produce gramophone records. Now used on cassettes.

Dynagroove. A method evolved by RCA for reducing tracing distortion on gramophone records.

Dynamic Range. The difference in volume between the quietest and the loudest passages of the music or other sound being played.

Equalization. The deliberate introduction of a particular frequency response into an amplifier to compensate for a non-flat response in the source of sound. For example, for technical reasons the frequency response produced during recording on disc or tape is not flat, so the amplifier which replays these recordings has to have a particular frequency response to compensate for this. Tone-control is a form of equalization.

Ergonomics. The science of the relationship between man and the machine. In this context it particularly applies to the use of controls and their functions.

Feedback. Taking part of the output of an amplifier and feeding it into the input. Negative feedback does this in a way that minimizes distortion, hum etc in the amplifier. (See "Acoustic Feedback.")

Filter. A circuit designed to remove or attenuate certain frequencies. In an amplifier filter circuits are often used to remove record hiss or scratch (low-pass filter) or turntable rumble (high-pass filter).

Flutter. A fluctuation in frequency of reproduced sound generally due to speed fluctuations in turntables or tape decks. Flutter refers generally to the faster fluctuations (say over 10 Hz). Lower-speed fluctuations are usually called "wow."

FM (Frequency Modulation). A method of applying an audio signal to a radio-frequency wave for transmission, so as to vary the carrier frequency. It is the method used by the BBC and others on their VHF sound transmissions. It allows good Hi-Fi quality reception with very little interference.

Frequency Response. The relative amplification of an amplifier or a complete audio system at various frequencies. If the amplification is the same at all stated frequencies then the frequency response is said to be flat.

Gain. The amount of amplification which occurs in an amplifier or system. If, when one volt of signal is put into an amplifier the output is 10 volts, then the gain is said to be 10 times. However, gain is often measured in decibels.

Harmonic Distortion. A form of distortion of a signal passed through an amplifier, which adds to the original sound harmonics or overtones of the original frequencies, ie if a 1000 Hz tone is put into the amplifier, the output would also be mainly a 1000 c/s tone but would have some amount of 2000 Hz, 3000 Hz etc added to it.

Heat Sink. A mass of metal, suitably shaped, on to which a transistor is bolted. Its purpose is to carry away the heat generated in the transistor, thus preventing failure of the transistor through overheating.

Hertz. The number of complete cycles per second executed by an alternating quantity. Named after Hertz, the discoverer of radio waves. Multiples are kilohertz, megahertz etc. Abbreviations are Hz, kHz, MHz.

High-pass Filter. A filter which will allow tones or musical notes above a certain frequency to pass but will block or attenuate lower tones. Such a filter can be used to minimize turntable rumble if it is designed to cut off frequencies below about 30 Hz.

Hum. An unwanted low-pitched sound produced in an amplifier from the AC mains or its harmonics; usually 50 Hz or its second harmonic 100 Hz. Can be caused by mains wiring or the mains transformer being too close to signal wiring, inadequate screening, inadequate smoothing in a power supply unit etc.

Infinite Baffle. (1) Ideally a flat board of infinitely large area with a loud-speaker mounted at the centre. (2) Generally used to mean a completely enclosed and sealed box, the only hole being that in which the loudspeaker is mounted. (See "Baffle.")

Integrated Amplifier. An amplifier in which the preamplifier, main amplifier and power supply are all contained in the one unit.

Integrated Circuit. An electronic package which includes a number of transistors, diodes, resistors and capacitors formed and connected on a common substrate by the use of modern microelectronic techniques. Abbreviation IC.

Intermediate Frequency (IF). In a superhet radio receiver this is the frequency to which the radio frequency is converted for amplification before detection.

Intermodulation. A form of distortion produced by imperfect amplifiers, loudspeakers, pick-ups etc, whereby the mixing of two different tones produces unwanted tones which are not harmonically related to either tone and are therefore very unpleasant.

Kilocycle (kilohertz). A kilohertz is equal to one thousand cycles per second. Hertz is now the preferred term to describe cycles per second, when referring to electrical or wave-propagated quantities which are met in audio and radio engineering.

Level. The loudness of reproduced sound or strength of signal.

Low-pass Filter. A filter which will allow tones or musical notes below a certain frequency to pass but will block or attenuate higher notes. Such a filter is often used to reduce unwanted noise from worn records (scratch filter).

Mass. The amount of matter involved in the part or body under consideration.

Matrix. An array of cross-connecting circuits used to encode or decode two- or four-channel stereo signals for recording or replay respectively. Quadraphonic programmes may thus be carried on two-channel systems such as gramophone records or stereo broadcasts. Different matrix systems have been adopted by various companies, eg the SQ, QS and RM matrices.

Megacycle (Mc/s). A megacycle per second is equal to a million cycles per second. Megahertz now replaces this term.

Micron. A millionth part of a metre, ie 10^{-6} metre or approximately $\frac{1}{25}$-mil.

Microsecond. A millionth part of a second.

Mil. 1 thousandth of an inch, ie 0·001 in.

Millisecond. A thousandth part of a second.

Mixer. An electronic circuit designed to combine and control several different signals so that the "mixed" or combined signal can be applied to a single amplifier.

Modulation. A method of superimposing a sound signal on to a radio frequency for transmission. (See "AM" and "FM.")

Mono (monophonic). Describes any system of sound reproduction which uses only one channel, ie either one loudspeaker only or, if more than one, then the same sound produced from each.

Motor-boating. A low-speed popping sound sometimes produced from a faulty or badly designed amplifier, even when no signal is being put into it.

Multiplex. A method of radio transmission of stereo signals, ie two separate signals, left and right sound, can be carried on one radio-frequency FM main channel by the use of an additional sub-carrier frequency. (See "Sub-carrier.") The CD-4 multiplex system is used on discs.

Negative Feedback (NFB). (See "Feedback.")

Objective. A measurement conducted by instrumental means which do not involve human decisions or judgments.

Oscillator. An electronic circuit which generates an alternating current. Used to produce tone signals in signal generators, electronic organs etc, and to produce radio frequencies where required in superhet receivers, radio transmitters etc.

Parameters. The fixed properties of a device which govern its performance.

PCM. An abbreviation for pulse code modulation, which is a digital method of transmitting programmes which has the advantage that the signal may be completely regenerated by repeaters without amplifying noise or deteriorating response.

Piezoelectric. The property of certain natural or synthetic crystals and ceramics whereby a voltage is generated when they are subjected to a mechanical force or strain.

Pinch Wheel. A rubber-tyred wheel holding the tape in contact with the capstan in a tape-recorder drive system.

Polar Response. The output of a microphone or loudspeaker plotted as a polar curve showing the variation at different angles relative to the main axis.

Potentiometer. (1) A simple attenuator using fixed resistors. (2) Often refers to the resistor with a continuously variable tapping point which is used in electronic circuits as volume, tone or other control. (See "Attenuator.")

Power Amplifier (or main amplifier) amplifies the required sound to a level suitable for connecting to the loudspeaker. It may be a separate unit for use with a preamplifier or it may be incorporated in a radio set, radiogram or integrated amplifier.

Preamplifier. In audio it has the necessary volume and tone controls and amplifies the small signals from pick-ups, tape-heads etc to a level suitable for feeding into the main amplifier. It is often convenient to keep the size of the preamp (control unit) small.

Precedence effect. A fundamental property of hearing whereby we assign direction to the first wavefront to arrive. Thus a sound source may be located in a complex sound field in a reverberant room. Originally discovered by De Haas.

Pressure. Acoustic pressure refers to the instantaneous variations of the atmospheric pressure above and below its static or undistured value caused by the passage of a sound wave.

Pressure Unit. A special type of loudspeaker designed to be mounted on a horn for production of sound. Often used in public-address work.

Quadraphony. This name has been given to 4-channel forms of stereophony in which front and rear loudspeakers are fed with appropriate signals derived from matrixing or multiplexing to give a better sense of depth and ambience than is provided by left- and right-hand stereo speakers only.

RF (Radio Frequency). Higher frequencies than audio. Used to carry sound signals in radio transmission.

Resonance. In electronic or mechanical systems it is a tendency for excessive response to a particular frequency. Bells or tuning forks are mechanical resonators. The tuned circuits in a radio receiver which give accurate selection of the wanted station display electrical resonance.

Reverberant Sound. The multiple reflected sound waves which persist in a room after a sound is originated.

Rumble. The low-frequency sound resulting from vibrations of the turntable being amplified.

Selectivity. The degree to which a radio tuner or receiver can reject stations or signals on frequencies close to the wanted one.

Semiconductor. An electronic device (such as a transistor or diode) which uses a solid material such as germanium or silicon as the medium instead of the vacuum used in valves. The properties of the devices depend largely on internal junctions which are formed between *p*- and *n*-type semi-conductor material.

Sensitivity. Describes the ability of a tuner or receiver to give reasonable reception of weak or distant stations. Generally given as the lowest number of microvolts of signal received which will give acceptable sound quality.

Signal-to-noise Ratio. The ratio of the wanted signal voltage to the voltage of unwanted signals (noise), generally expressed in dB. The higher the signal-to-noise ratio the lower will be the background noise on the programme.

Solid State. Describes circuits which use semi-conductors (transistors etc) rather than valves.

Squelch. Circuits on FM sets which automatically silence the inter-station noise received while tuning the receiver.

Stereophony. Dual or multi-channel techniques for giving sound-directional information with a potentially considerable increase in realism.

Stiffness. The reciprocal of compliance. Is normally defined as a mechanical spring constant.

Sub-carrier. An additional carrier wave transmitted in the FM stereo multiplex system which conveys the stereo channel difference information, the sum being conveyed by the main carrier wave.

Subjective. A personal judgment or impression which is used to give a quality or other rating of sound reproduction etc.

Tape-recorder. A magnetic tape-recorder used for sound or vision recording; single or multi-channel recording is possible.

Transducer. An energy-converting device such as a microphone, pick-up or loudspeaker which interchanges acoustical, mechanical or electrical signals.

Transient Response. The ability of a loudspeaker or amplifier to handle sudden bursts of sound such as when a cymbal or drum is struck, or when a string is plucked.

Treble. The high notes or frequencies in the audio range.

Tweeter. A loudspeaker designed specifically for reproducing high notes or high frequencies.

UHF (Ultra High Frequency). Radio frequency greater than 300 megacycles (higher than VHF). BBC and ITV transmit on UHF radio waves.

Varactors. Special semi-conductor diodes which are designed to provide given changes in capacity when the DC bias is changed, thus enabling them to be used as radio tuning devices.

VHF (Very High Frequency). Radio frequency between 30 megacycles and 300 megacycles. BBC 1, ITV and BBC FM sound use VHF radio waves.

Wave. A propagated travelling disturbance of sound or any other disturbance such as radio waves.

Wavefront. The salient face of an advancing wave.

Woofer. A loudspeaker designed specifically for reproducing the low notes or low frequencies.

Wow. Slow fluctuations in the speed of a turntable or tape-deck. (See also "Flutter.")

Note. The abbreviation for grammes weight is g. The gravitational acceleration on a falling body is represented by italic g. In most cases the distinction may be inferred by the context.

Recording and copyright

Any person making a recording of a record or a programme, broadcast or otherwise, should take notice of the laws of copyright. Many people have assumed that home copying of commercial records etc for private use is permissible, but it is in fact an infringement of the laws of copyright, unless permission has been granted.

In order to rationalize the position, the Mechanical Copyright Protection Society, 380 Streatham High Road, London SW16, has arranged to issue a licence for the sum of £1.50 + VAT which allows copying of discs on to tape for private home listening by the individual. Other licences may be issued by this Society for charitable or semi-private purposes, such as tape-recording clubs and competitions. The Federation of British Tape Recordists and Clubs, c/o Mr John Bradley, 33 Fairlawns, Malden Road, Wallington, Surrey, can supply more information and may assist members to obtain favourable terms.

The above obviously represent very fair arrangements, and all amateur recordists wishing to copy discs and programmes should certainly avail themselves of a licence arrangement. It not only protects them against copyright infringement actions, but it also helps artists and recording companies to obtain a fair return for their work.

Recommended books for further reading

[1] *Experiments in hearing.* G. Von Békésy (McGraw-Hill 1960).

[2] *Acoustical techniques and transducers.* M. L. Gayford (Macdonald and Evans 1962). Deals in detail with the design of microphones, pick-ups, loudspeakers and the acoustic treatment of rooms and studios.

[3] *Hi-Fi in the Home.* John Crabbe (Blandford Press 1969). An excellent and readable treatment of the whole subject.

[4] *Aerials Handbook.* G. A. Briggs (Wharfedale Wireless Works. 2nd Edition, 1968). A useful survey of high-quality aerial systems.

[5] *FM simplified.* Milton S. Kiver (Van Nostrand, 3rd edition, 1960). A clear and thorough exposition of all aspects of modern FM transmission and reception, including explanations of "capture effect" etc.

[6] *Transistor Receivers and Amplifiers.* F. G. Rayer (Focal Press, 1966). A useful and representative collection of circuits for radio sets, FM tuners, pre-amplifiers and power-amplifiers.

[7] *Pick-ups the Key to Hi-Fi.* J. Walton (Pitman. 2nd edition, 1965). A clear and readable exposition by an expert of the complex phenomena involved.

[8] *Tape Recording and Reproduction.* A McWilliams (Focal Press, 1964). A thorough introduction to tape-recording theory, mechanisms and electronics.

[9] *Sound and Ciné for Beginners.* R. Golding (Miles Henslow Books, 1965).

[10] *Magnetic Recording for the Hobbyist.* (Foulsham—Sam's Technical Books.)

[11] "Stereo Gramophone Pickups—a review of transducer types." Stanley Kelly. (*Wireless World*, December 1969).

[12] "The Construction of a Gimbal Bearing Pickup Arm with Bias Compensation." (*Hi-Fi News*. February 1970.)

[13] *Manual of Sound Recording.* John Aldred (Fountain Press). A comprehensive basic survey of recording amplifiers, stereo, loudspeakers etc.

[14] *High Quality Sound Reproduction.* James Moir (Chapman & Hall, 1961). Detailed description of microphones, recorders, loudspeakers and considerable data on amplifier and tone control design etc.

[15] "Mullard Amplifier Designs." (On application to Mullard Ltd.) A series of data sheets issued at intervals by Mullard Ltd giving full design details of

valve and transistor preamplifiers, power-amplifiers, tape-recorder electronics etc. Also *Mullard Transistor audio circuits book* (1970) £1.50.

[16] *Loudspeakers.* E. J. Jordan (Focal Press, 1963). A readable professional survey of loudspeaker and cabinet theory and practice.

[17] *Loudspeakers in your home.* Ralph West (Miles Henslow Hi-Fi Books, 1966). A very readable up-to-date factual survey by a well-known expert of modern loudspeakers of all kinds.

[18] *Cabinet Handbook.* G. A. Briggs (Wharfedale Wireless Works 1962). The design and construction of many types of loudspeaker cabinets.

[19] *Goodmans High Fidelity Loudspeaker Manual.* (Free on request from Goodmans Ltd, Wembley, Middlesex.) Details of all Goodmans loudspeakers and suitable cabinet designs.

[20] "Home Cabinet Design and Construction." (*Hi-Fi News*, June 1970.) A useful article giving ideas on good styling and building of cabinets for amateurs.

[21] *Loudspeakers.* G. A. Briggs (Wharfedale Wireless Works, 5th edition, reprinted 1969). A comprehensive practical book on loudspeakers and cabinets.

[22] *Tape-Recorder Manual* C. W. Hellyer (Newnes). A useful volume covering the operation, circuits and servicing procedures for most popular tape-recorders in use.

[23] *FM Radio Servicing Handbook.* Gordon J. King. (Butterworth, 2nd edition 1970.) A very sound straightforward treatment of modern FM and stereo receiver techniques for service engineers and amateurs.

[24] *Living with Hi-Fi.* John Borwick. (General gramophone publications, Harrow, 1972.) A good general introduction to all aspects of Hi-Fi and equipment. Also *Quadraphony now*, booklet, 45p, 1974.

[25] *Studio microphone technique.* Michael Thorne. (*Studio Sound*, July 1973.) This article deals especially with close microphone pick-up of musical instruments in smaller rooms and studios.

[26] *The design and construction of high-quality loudspeakers and cabinets.* Leaflets A 60 to A 64. Prices on application to Radford Acoustics, Bristol. Designs and drawings for cabinets and use of kits for high-quality omni and forward facing cabinet loudspeakers. Highly recommended.

[27] Articles on SQ and other decoder circuits: *Electronic Engineering*, p. 49, November 1973. Also Geoffrey Shorter, *Wireless World*, p. 114, March 1973.

[28] *Horn loaded loudspeakers.* Paul W. Klipsch. *Wireless World*, p. 50, February 1970. Article on advantages of modern horn designs and details of folded horn design. Also *W.W.*, March/April/May 1974. Articles by T. Dinsdale.

Hi-Fi publications and services

Hi-Fi News and Hi-Fi Year Book
Studio Sound and Tape recorder
Hi-Fi Annual
Wireless World

All the above published by Link House Publications Ltd, Dingwall Avenue, Croydon CR9 2TA and IPC Electrical-Electronic Year Books Ltd, Dorset House, Stamford Street, London SE1

Hi-Fi Sound
Hi-Fi Sound Annual
Hi-Fi Answers
Popular Hi-Fi

Published by Haymarket Publishing Group, Gillow House, 5 Winsley Street, London W1

The Gramophone

107–9 Kenton Road, Harrow, Middlesex HA3 0HA

Audio and Record Review

Hanover Press, Mill Street, Hanover Square, London W1

"Record Information and Review." A monthly Newsletter from EMG Handmade Gramophones Ltd, 26 Soho Square, London W1V 6BB

The Wilson Stereo Library Ltd. Comprehensive record index and sales etc. 104–6 Norwood High Street, London SE27

Audio (Hi-Fi magazine)

IPC Magazines Ltd, Fleetway House, Farringdon Street, London EC4A 4AD

Hi-Fi for Pleasure

Blakeham Productions Ltd, 23 Denmark Street, London WC2H 8NA

German periodicals dealing with Hi-Fi and electronics

Hi-Fi Stereophonie
FUNKSCHAU, Radio, Hi-Fi, Electronics
Publisher: Franzis-Verlag, München

Hi-Fi for the enthusiast

Radio Mentor
Publisher: Radio Mentor-Verlag, Berlin

Fono-Forum

British Standards on amplifiers, tuners etc, British Standards Institution, London.
German Standards for industry (Domestic and Hi-Fi equipment):
DIN 45511 and DIN 45500.

High Fidelity Dealers Association (HFDA), 19–21 Conway Street, Fitzroy Square, London W1P 6DY

Gramophone test records

These are sold by some of the major gramophone-record makers and are valuable for assessing the performance of pick-ups on frequencies of specified levels and also on musical excerpts of particular severity and interest, eg—

Decca/Polydor Decca House, London SE1 7SW

EMI Records TCS 101, 102, 104, 105. RIAA frequency recordings (also test tapes for tape-recorders, CCITT frequency recordings). EMI, Hayes, Middlesex. Also TCS 201.

Further test records
Decca LXT 5346. Frequency bands
Decca SKL 2057. Frequency bands
Shure TTR-101. Stereo phasing and trackability tests.
Hi-Fi Stereo Review 211 Stereo balance, phasing, test tones. (Available from Wilson Stereo Library, London.)
CBS STR 120, 140, 160. Frequency, noise bands, vertical tracking, etc.
CBS STR 111. Square wave tests, tracking, intermodulation, etc.
Audix ADX 301. Frequency bands, musical tests.
CBS STR 100. Frequency bands, stereo separation, stylus checks, compliance tests.
Shure "audio obstacle course and tracking tests" TTR 110.
Shure TTR 103. New test record, used with advantage with an oscilloscope. (Shure records obtained from agents and Shure Ltd, Eccleston Road, Maidstone, ME15 6AU.)
Hi-Fi Sound test record HFS 69. (From Haymarket Publishing Co, 34 Fouberts Place, London, W1 or Howland West Ltd, London, NW3.)
This record includes 10 groups of tests including white noise, tone bursts and music, many being designed to give positive stereo checks without the use of instruments.
Fisher Radio, CD-4 and SQ performance test record (from Hayden Laboratories Ltd, Amersham, Bucks).
Bruel & Kjaer QR 2009, QR 2010 cartridge test records. QR 2011 1/3 octave pink noise record.
Slide synchronizer. The Philips N6400 slide synchronizer unit will operate an automatic slide projector from cues provided by a tape-recorded programme.

Hi-Fi for the enthusiast

Guide to record care and cleaning

Booklet "Guide to Better Care of LP and Stereo Records." Obtained from
C. E. Watts Ltd, Darby House, Sunbury-on-Thames, Middlesex

Further record and tape accessories and cleaning materials, attachments etc:
Colton Musonic Ltd, St Albans, Hertfordshire
Metrocare Accessories. Metrosound Audio Products Ltd, Waltham Abbey,
Essex
Bib Hi Fi Accessories, Hemel Hempstead, Herts
The "Zerostat" piezoelectric gun, Zerostat, St Ives, Hunts

Index